Mathematical Sciences Research Institute
Publications

24

Editors

S.S. Chern
I. Kaplansky
C.C. Moore
I.M. Singer

Mathematical Sciences Research Institute
Publications

S. Montgomery   L. Small
Editors

Noncommutative Rings

Springer-Verlag
New York Berlin Heidelberg London Paris
Tokyo Hong Kong Barcelona Budapest

Susan Montgomery
Department of Mathematics
University of Southern California
Los Angeles, CA 90089
USA

Lance Small
Department of Mathematics
University of California at San Diego
La Jolla, CA 92093
USA

Mathematical Sciences Research Institute
1000 Centennial Drive
Berkeley, CA 94720
USA

The Mathematical Sciences Research Institute wishes to acknowledge support by the National Science Foundation.

Mathematics Subject Classifications: 16-00, 16A33, 16A27

Library of Congress Cataloging-in-Publication Data
Noncommutative rings / Susan Montgomery, Lance Small, editors.
    p. cm. — (Mathematical Sciences Research Institute
publications ; 24)
    Lectures delivered at the Microprogram on Noncommutative Rings,
held at MSRI, July 10–21, 1989.
    Includes bibliographical references.
    ISBN 0-387-97704-X. — ISBN 3-540-97704-X
    1. Noncommutative rings — Congresses. I. Montgomery, Susan.
II. Small, Lance W., 1941– . III. Mathematical Sciences Research
Institute (Berkeley, Calif.) IV. Microprogram on Noncommutative
Rings (1989 : Mathematical Sciences Research Institute) V. Series.
QA251.4.N66 1991
512′.4 — dc20   91-32480

Printed on acid-free paper.

Photocomposed copy prepared by the Mathematical Sciences Research Institute using TEX.
Printed and bound by Edwards Brothers, Inc., Ann Arbor, MI.
Printed in the United States of America.

9 8 7 6 5 4 3 2 1

ISBN 0-387-97704-X Springer-Verlag New York Berlin Heidelberg
ISBN 3-540-97704-X Springer-Verlag Berlin Heidelberg New York

This volume is dedicated to the memory of
Robert B. Warfield

Preface

This volume collects some of the survey lectures delivered at the Micro-program on Noncommutative Rings held at MSRI, July 10–21, 1989. While the program was concerned with recent advances in ring theory, it also had as an important component lectures on related areas of mathematics where ring theory might be expected to have an impact.

Thus, there are lectures of S. P. Smith on quantum groups and Marc Rieffel on algebraic aspects of quantum field theory. Martin Lorenz and Donald Passman consider in their lectures various aspects of crossed products: homological and K-theoretic to group actions. Kenneth Brown presents the "modern" theory of Noetherian rings and localization.

These contributions as well as the others not presented here show that ring theory remains a vigorous and useful area.

The planning and organization of the program were done by the undersigned and the late Robert Warfield. His illness prevented his attendance at the meeting. It is to him we dedicate this volume.

The organizers wish to extend their thanks to Irving Kaplansky, Director of MSRI, and the staff for all of their efforts in making this conference such a success.

Susan Montgomery
Lance Small

NONCOMMUTATIVE RINGS

TABLE OF CONTENTS

The Representation Theory of Noetherian Rings

KENNETH A. BROWN

Dedicated to my friend and colleague Bob Warfield

Introduction

The purpose of this article is to give a survey of part of what is known about the structure of the indecomposable injective modules over a Noetherian ring R, to indicate how this structure depends on the nature of certain bimodules within the ring which afford "links" between the prime ideals of R, and to suggest some directions for future research in these areas.

The subject began with Matlis' analysis of the commutative case in 1958 [**31**]. During the sixties and beyond many ring theorists studied the Artin-Rees property in various classes of noncommutative Noetherian rings, at least partly in an attempt to extend Matlis' theory to such rings. The impossibility of this endeavour was already becoming apparent by the early seventies, notably through Jategaonkar's papers [**20, 21**] which first highlighted the connections between bimodules, localization at prime ideals and the structure of injective modules. His insights stimulated much subsequent research, focussed largely on the search for a useful and workable theory of localization "near" a prime ideal. But inevitably—in view of the connections mentioned above—this work also shed light on representation theory, and in recent years the representation theoretic point of view has assumed a more central position.

Since this entire paper is in effect no more than an extended introduction, (in which those proofs included are designed to be "skippable"), I won't discuss the contents in detail here. In outline, they are as follows. Noetherian bimodules and their connection with representation theory via Jategaonkar's "Main Lemma" form the subject of §1. In §2 I discuss various open questions, and a few results, concerning those properties which can be passed from one side of a Noetherian bimodule to the other. With §3 we begin the study of indecomposable injective modules over a Noetherian ring. The emphasis in this section is on rings with the right strong second layer condition (Definition 5); in §4 I describe recent work which aims to encompass all Noetherian rings within the framework developed in §3. Finally, we specialise to Noetherian PI-rings in §5.

Throughout, I have tried to highlight some of the many open questions in this area; in fact I've listed 15 of these at appropriate points in the text.

Most of the results in this paper have appeared elsewhere, the chief exception being Theorem 10, a result of K.R. Goodearl on bimodules over simple rings, which answers a question I raised in my lectures at the MSRI conference. I have tried to include full references to the literature, especially where I have not included the full proof of a result. For the reader who wishes to take the subject further, the relevant books are [18], [25], and [32]. With such a reader in mind I have included references to some papers which, though related to the subject at hand, are not mentioned in the text; but I am afraid the list of references should not be regarded as exhaustive in this respect. And the text itself is certainly very far from exhaustive. In particular, to escape drowning in a sea of technicalities, I have not discussed what extra information is available for specific classes of rings, with the (brief) exception of PI-rings in §5. The most noteworthy absentee in this respect is the class of enveloping algebras of solvable Lie algebras, for which detailed results can be found in [11, 12].

ACKNOWLEDGEMENTS: Some of the work for this paper was done while I was visiting the University of Washington (Seattle) in the summer of 1989, with financial support from the N.S.F. and the Carnegie Trust for the Universities of Scotland.

§1. Bimodules

DEFINITION 1: (i) A *bond* B between prime Noetherian rings S and T is a non-zero S-T-bimodule which is finitely generated and torsion-free both as a left S-module and as a right T-module.

(ii) Let R be a Noetherian ring with prime ideals P and Q. There is an *internal bond*, or *ideal link*, from Q to P if there exist ideals A and B of R with $A \subset B$ and B/A a torsion free left R/Q-module and a torsion-free right R/P-module. In this case we write $Q \approx P$.

(iii) If $Q \approx P$ as in (ii) with $QP \subseteq A$, we say that Q is *(right second layer) linked to* P, or that there is a *(second layer) link from* Q *to* P, and write $Q \rightsquigarrow P$.

REMARK 2: (i) If $Q \rightsquigarrow P$ then it is easy to see that there is an internal bond of the form $Q \cap P/A'$; thus second layer links are internal bonds occurring 'as high as possible' in R.

(ii) The example $R = \begin{pmatrix} \mathbb{C} & \mathbb{C} \\ 0 & \mathbb{C} \end{pmatrix}$, $Q = \begin{pmatrix} 0 & \mathbb{C} \\ 0 & \mathbb{C} \end{pmatrix}$, $P = \begin{pmatrix} \mathbb{C} & \mathbb{C} \\ 0 & 0 \end{pmatrix}$ shows that the definitions 1(ii) are not symmetric.

(iii) It is a consequence of the definitions that $Q \approx P$ if $Q \rightsquigarrow\!\!\rightarrow P$. The converse is false: In the ring $R = \begin{pmatrix} \mathbb{C} & \mathbb{C} & \mathbb{C} \\ 0 & \mathbb{C} & \mathbb{C} \\ 0 & 0 & \mathbb{C} \end{pmatrix}$, $Q = \begin{pmatrix} 0 & \mathbb{C} & \mathbb{C} \\ 0 & \mathbb{C} & \mathbb{C} \\ 0 & 0 & \mathbb{C} \end{pmatrix} \approx \begin{pmatrix} \mathbb{C} & \mathbb{C} & \mathbb{C} \\ 0 & \mathbb{C} & \mathbb{C} \\ 0 & 0 & 0 \end{pmatrix} = P$ via the ideal link $\begin{pmatrix} 0 & 0 & \mathbb{C} \\ 0 & 0 & 0 \\ 0 & 0 & 0 \end{pmatrix}$, but $P \cap Q = QP$. Nevertheless, for rings with the second layer condition (see Definition 5 below) $\rightsquigarrow\!\!\rightarrow$ and $\approx$ generate the same equivalence relation on $\mathrm{Spec}(R)$. (For this, see [25, Theorem 8.2.4], or consult Theorem 26 below.) This is no longer true for rings without the s.l.c. [39, Theorem 4.1].

(iv) The importance of internal bonds for representation theory was first pointed out in [20]. Second layer links and the notation $Q \rightsquigarrow\!\!\rightarrow P$ were introduced (for fully bounded Noetherian rings) in [33]. The notation $Q \approx P$ was first used in [28].

The first connections between the ideal structure of a Noetherian ring and its representation theory are achieved through what has come to be called Jategaonkar's Main Lemma. To state it we need one more set of definitions.

DEFINITION 3: Let Γ be a subset of $\mathrm{Spec}(R)$. An ideal I of R is Γ-*semiprime* if and only if I is a finite intersection of primes in Γ. An R-module M is Γ-*semiprime* if and only if the annihilator of every finitely generated submodule of M is Γ-*semiprime*. If Γ is finite and $S = \cap\{P : P \in \Gamma\}$, we shall also say that M is S-*semiprime*. If $\Gamma = \{S\}$, we say that M is S-*prime*.

LEMMA 4. (i) (Jategaonkar [25, 6.1.2, 6.1.3]) *Let P and Q be prime ideals of R. Let*

(1) $$0 \longrightarrow U \longrightarrow M \longrightarrow V \longrightarrow 0$$

be an exact sequence of R-modules, with U, M and V all finitely generated, non-zero and uniform. Suppose that $(\alpha)U$ is P-prime, $(\beta)V$ is Q-prime and (γ) for all submodules M' of M not contained in U, $\mathrm{Ann}(M') = \mathrm{Ann}(M) = A$. Then either

  (a) $Q \rightsquigarrow\!\!\rightarrow P$ *via the second layer link* $Q \cap P/A$, *or*
  (b) $A = Q \subsetneq P$.

If (a) holds and U is (R/P)-torsion-free, (so that U is a right ideal of R/P), then V is (R/Q)-torsion-free, (so that V is a right ideal of R/Q). If (b) holds, then V is (R/Q)-torsion.

(ii) [**18**, Theorem 11.2] *Let P and Q be prime ideals of R, and suppose that* $Q \rightsquigarrow\rightsquigarrow\rightarrow P$. *Then there exists an exact sequence of R-modules satisfying all the hypotheses of (1) in (i), with U R/P-torsion free and V R/Q-torsion free.*

Since it is an archetype for many proofs in this area, let us consider the

PROOF OF (i): Clearly $QP \subseteq A$. Suppose that $A \subsetneq Q \cap P$ and let I be an ideal with $A \subsetneq I \subseteq Q \cap P$. Since $0 \neq MI \subseteq U$, $r - \text{Ann}(I/A) = P$ by (α). Hence $(Q \cap P)/A$ is a non-zero torsion free right R/P-module. Let $T = l - \text{Ann}(I/A)$, and suppose that $Q \subsetneq T$. Then $MT \not\subseteq U$ by (β), so $\text{Ann}(MT) = A$ by (γ). But $(MT)I = 0$, so $I \subseteq A$, a contradiction. Therefore $(Q \cap P)/A$ is a torsion free left R/Q-module, and so $Q \rightsquigarrow\rightsquigarrow\rightarrow P$ via $(Q \cap P)/A$.

Suppose on the other hand that $A = Q \cap P$, so that $MPQ = 0$. Now $MP \neq 0$ since $U \neq M$, so that $MP \cap U \neq 0$ because M is uniform, and so $Q \subseteq P$ by (α). Thus $A = Q$, and, (again using $U \neq M$), this ideal is strictly inside P.

Finally, suppose that (a) holds and that U is R/P-torsion free. Let $m \in M$ and $d \in R$, d regular modulo Q, with $md \in U$. Since the right R-endomorphism of $Q \cap P/A$ afforded by left multiplication by d is a monomorphism, $d(Q \cap P) + A/A$ is right essential in $(Q \cap P)/A$. Hence $(Q \cap P)/\{d(Q \cap P) + A\}$ is a torsion right R/P-module. The homomorphism $R \longrightarrow M : r \longrightarrow mr$ contains $d(Q \cap P) + A$ in its kernel, so that $m(Q \cap P)$ is a torsion R/P-submodule of U. By hypothesis, $m(Q \cap P) = 0$, so $m \in U$ by (γ) and (a). That is, M/U is R/Q-torsion free.

Consider the content of Lemma 4 in two particular cases. First, *suppose that R is commutative.* Then (b) cannot occur, as there are no finitely generated faithful torsion modules over a commutative domain. And when R is commutative, (or even when $Q \cap P$ is generated by central elements), $Q \rightsquigarrow\rightsquigarrow\rightarrow P$ only if $Q = P$. So in any exact sequence of form (1), U and V are right ideals of the same prime factor of R. This is the essence of the Artin-Rees property [**18**, Lemma 11.11].

Second, let $S = A_1(\mathbb{C}) = \mathbb{C}[x, y : xy - yx = 1]$, the first Weyl algebra, and let $R = \mathbb{C} + xS$, the idealizer in S of xS, so R is a Noetherian domain with unique proper ideal $P = xS$ [**32**, Theorem 5.5.10]. There are thus no links between the two primes of R (noting $P^2 = P$), so only (b) of Lemma 4 can happen. In fact case (b) does occur: The R-module $M = S/xS$ is of this form, with $U = R/xS$ and $Q = 0$ [**32**, Theorem 1.1.12(ii)].

DEFINITION 5: (i) A prime ideal P of R satisfies the *(right) strong second layer condition (r.s.s.l.c.)* if only (a) of Lemma 4 can occur for P. The same definition is extended to subsets of $\mathrm{Spec}(R)$, and to R itself, (so that R satisfies the r.s.s.l.c. exactly when all its prime ideals do).

(ii) If, for given P and Q, there exists an exact sequence as in case (b) of Lemma 4, Q is said to be (right) *jump linked to* P. We shall denote this by: $Q \angle P$.

Conditions closely related to the strong second layer condition were studied independently by myself and by Jategaonkar in the late seventies, building on earlier work of Jategaonkar and Mueller [20, 21], [33]. We obtained results reported as [7] and [22], which later appeared in [8] and [23, 24]. (To be precise, [8] examines *right smooth rings*, which are the same as rings with the r.s.s.l.c. in circumstances where the Bimodule Conjecture— Question D below—is true. The condition (*) studied in [22] is now called the *right second layer condition*; it is (formally, at least) weaker than the r.s.s.l.c. The present terminology (apart from Definition 5(ii)) was finalised in [25].

EXAMPLES 6: (i) Just as with commutative rings, the absence of faithful finitely generated torsion modules over prime factor rings ensures that fully bounded Noetherian rings (and hence in particular Noetherian rings satisfying a polynomial identity) satisfy the r.s.s.l.c. [18, Lemma 8.2].

(ii) Group rings (and even crossed products) of polycyclic-by-finite groups over commutative Noetherian coefficient rings satisfy the r.s.s.l.c.[8], [24], [2], [25, Theorem A.4.6].

(iii) The enveloping algebra $U(\mathbf{g})$ of a finite dimensional Lie algebra $\mathbf{g}$ over a field K has the r.s.s.l.c. if and only if $\mathbf{g}$ is solvable or K has positive characteristic [8], [22], [25, Theorem A.3.9].

(iv) Matrix rings over, and factor rings of, rings with the r.s.s.l.c. retain the property.

(v) If R is a Noetherian subring of a ring S such that $_R S$ and S_R are finitely generated then R has the r.s.s.l.c. only if S does too [30].

The fact (iii) that non-solvable Lie algebras in characteristic zero have enveloping algebras which fail to satisfy the r.s.s.l.c. can be read off from the next lemma and basic properties of simple Lie algebras [16].

DEFINITION 7: The *right clique* of a prime P of R is $r.cl(P) = \{Q \in \mathrm{Spec}(R) : Q = P_n \rightsquigarrow P_{n-1} \rightsquigarrow \cdots \rightsquigarrow P_1 = P,$ for some $n\}$. The *clique* of P, $cl(P)$, is obtained by permitting the arrow between each pair P_i, P_{i-1} of primes above to point in either direction.

Thus, for example $R = \begin{pmatrix} \mathbb{C} & \mathbb{C} \\ 0 & \mathbb{C} \end{pmatrix}$ following Definition 1, $r.cl(Q) = \{Q\}$, whereas $cl(Q) = \{Q, P\}$. In general, $|r.cl(P)|$ is infinite; for example, when

$R = \mathbb{C}[x, y: xy - yx = y]$, the enveloping algebra of the two-dimensional solvable non-Abelian Lie algebra, $r.cl(\langle x, y \rangle) = \{\langle x - n, y \rangle : n \geq 0, n \in Z\}$ [**18**, Exercise 11G], [**25**, Example 5.4.9]. But cliques are always countable [**40**], [**18**, Theorem 14.23].

LEMMA 8. *Suppose that n is a positive integer and that R has a prime factor ring T with exactly n proper ideals. Then R does not satisfy the r.s.s.l.c. .*

PROOF: We may assume that $T = R$, in view of Examples 6(iv). Let P be a proper prime ideal of R, and set $I(P) = \cap\{Q: Q \in cl(P)\}$ and $J = J(P) = \cap\{I(P)^j : j \geq 0\}$. Thus J is a non-zero ideal, since R is prime. Let $E = E_R(R/P)$. Then J annihilates every submodule of E built by a sequence of extensions of type (a) from Lemma 4. But non-zero injective modules over prime Noetherian rings are faithful, (as is easily deduced from the existence of a regular element in every non-zero ideal [**32**, Proposition 2.3.5(ii)]). Thus there exist type (b) extensions in E.

We end this section with three sample questions among many which could be asked about the s.s.l.c. Part of the problem here is the paucity of examples; at present all known rings failing to have the s.s.l.c. fall into the pattern of the above lemma, but this must surely just be a failure of imagination.

QUESTION A. *Does every local ring satisfy s.s.l.c.?*

By a *local ring* we mean a Noetherian ring whose factor by its Jacobson radical is simple Artinian. Counterexamples exist among rings which are only *right* Noetherian [**19**]. It might be easier to determine whether the Jacobson radical of such a ring has the s.s.l.c., but of course this is equivalent to the old chestnut which asks whether this ideal satisfies the Artin-Rees property [**18**, Question 2, p. 285]. As is well-known, a positive answer would settle Jacobson's conjecture in the affirmative for local rings. But why stick to local rings:

QUESTION B. *Do the prime ideals minimal over the Jacobson radical of a Noetherian ring satisfy the s.s.l.c.?*

I suspect that the absence of counterexamples to the final question of this set is only temporary. The notation is as introduced in the proof of Lemma 8.

QUESTION C. *Let P be a prime ideal of R. Does $P/I(P)$ satisfy the s.s.l.c. in $R/I(P)$? How about $P/J(P)$ in $R/J(P)$?*[1]

---

[1]See note added in proof, page 24.

§2. Invariants of bonds

The questions I want to record in this short section may appear rather tangential to the main topic of this paper, but our ignorance on these matters leads to technical problems in so many proofs and definitions that their inclusion seems unavoidable. The first of the variants listed below is sometimes called the Bimodule Conjecture; it is maybe not surprising that no-one's surname attaches to this, since most bets seem to be laid on its having a negative answer. The *Krull dimension* referred to here is that of Gabrièl and Rentschler [**32**, Chapter 6] [**18**, Chapter 13].

QUESTION D. *If $_SB_T$ is a bond, is the left Krull dimension of S equal to the right Krull dimension of T?*

This is open even when $S = B = T$. By a result of Lenagan [**27**], [**18**, Theorem 7.10] the answer is "yes" when S or T is Artinian.

QUESTION E. *Do there exist internal bonds with $Q \subsetneq P$?*

QUESTION F. *Do there exist second layer links with $Q \subsetneq P$?*

Of course, a positive answer to Question D would force negative answers for Questions E and F. More generally, suppose there is a dimension function d defined on the isomorphism classes of finitely generated right and left R-modules such that d is (i) exact, meaning that if N and M are finitely generated R-modules with $N \subseteq M$ then $d(M) = \max\{d(N), d(M/N)\}$, (ii) *symmetric*, meaning that $d(_RB) = d(B_R)$ for every Noetherian R-R-bimodule B, and (iii) *prime separating*, meaning that $d(R/Q) > d(R/P)$ whenever Q and P are primes of R with $Q \subset P$. If B is an R/Q-R/P-bond then (i) and (ii) force $d(R/Q) = d(R/P)$, so (iii) yields negative answers for Questions E and F. When R is the enveloping algebra of a finite dimensional Lie algebra the Gel'fand-Kirillov dimension satisfies these three conditions [**26**], and so Questions E and F have negative answers in this case.

For fully bounded Noetherian rings the (Gabrièl-Rentschler) Krull dimension coincides with the classical Krull dimension defined in terms of chains of prime ideals [**18**, Theorem 13.13], so the following result settles Question D when S and T are fully bounded.

THEOREM 9. (Jategaonkar [**25**, Theorem 8.2.8], [**18**, Corollary 12.5]). *Let $_SB_T$ be a bond and suppose that S and T have the s.s.l.c. Then the classical Krull dimensions of S and T coincide.*

To prove Theorem 9, one finds a subfactor of B bonding a factor of S to a factor of T, and invokes induction. The result and proof are valid in the (formally weaker) setting where the adjective "strong" is omitted; but the example discussed after Lemma 4, $S = A_1(\mathbb{C}) = \mathbb{C}[x, y]$, $T = \mathbb{C} + xS$, $B = xS$ shows that some such hypothesis is needed.

Of course the theme of the above questions can (and should) be continued, by asking which other properties of prime rings can be passed across bonds and links. As one example among many, we may ask for "homological invariants" of bonds. But there are problems here. For example, since every prime Noetherian ring which is a finite module over a commutative domain is bonded to that ring, one cannot hope for much along these lines without imposing further constraints on the rings involved.

QUESTION G. *Let $_SB_T$ be a bond. Under what circumstances is B_T a projective module?*

At the MSRI conference I suggested that this might be so if S and T are simple rings, and within twenty-four hours this was confirmed by K.R. Goodearl. With Goodearl's permission I end this section with a proof of his result—it is an elegant illustration of the techniques which have been developed for studying Noetherian bimodules.

THEOREM 10. *([Goodearl, 1989]). Let $_SB_T$ be a bond between the (prime Noetherian) rings S and T. Suppose that S is simple. Then B_T is projective.*

PROOF: If R is a ring and M a right R-module, denote the dual module $\operatorname{Hom}_R(M, R)$ by M^*. We begin with three well-known but very useful facts.

(i) *Let T be a prime Noetherian ring and let M_T be a finitely generated torsion-free module. If the endomorphism ring E of M_T is simple, then M_T is projective.*

Proof. Since MM^* is an ideal of E, it is either 0 or E. Suppose that $MM^* = 0$. Then $0 = (MM^*)M = M(M^*M)$, so the ideal M^*M of T must be 0, since M_T is torsion-free. This forces M^* to be zero, which is impossible since M is finitely generated. Therefore $MM^* = E$, and so by the Dual Basis Lemma [**32**, Lemma 3.5.2] M_T is projective.

(ii) *Let $_SM_R$ be a bimodule, finitely generated on each side. Suppose that S is prime Noetherian and that $_SM$ is torsion. Then there is a non-zero ideal I of S with $IM = 0$.*

Proof. Let $m_1, \ldots, m_n$ be elements of M with $M = \Sigma_i m_i R$. Then $\cap_i \{s \in S: sm_i = 0\}$ annihilates M, and it's non-zero, as it is a finite intersection of essential left ideals of S.

(iii) Let S, T and B be as in the theorem, with S simple. Then there is a finite chain of bimodules $0 \subset B_1 \subset \cdots \subset B_n = B$ such that each factor B_i/B_{i-1} is an S-T-bond and a simple bimodule.

Proof. Choose a non-zero sub-bimodule C of B with $u - \dim(_S C)$ as small as possible. Let D be any non-zero sub-bimodule of C. Then $_S(C/D)$ is torsion, by our choice of C. Now (ii) shows that $D = C$, so C is simple. Now let F/C be the T-torsion submodule of B/C. Note that F/C is a bimodule, so by (ii) again there is a non-zero ideal I of T with $FI \subseteq C$. Let x be a regular element of T contained in I, so that $_S F \cong {}_S Fx \subseteq C$. Hence $u - \dim(_S F) = u - \dim(_S C)$, so $_S(F/C)$ is torsion; and the same argument as before shows that $F = C$. Thus B/C satisfies the same hypotheses as B, and (iii) follows by the ascending chain condition.

Now we proceed to the proof of the theorem. By (iii), we may assume that B is a simple bimodule. Set $E = \text{End}(B_T)$, so E contains S. We prove next that $_S E$ is finitely generated. Let $b_1, \ldots, b_n$ generate B as a T-module. Then there is an embedding as left S-modules of E in $B^{(n)}$ given by $e \longrightarrow (eb_1, \ldots, eb_n)$, for $e \in E$. Thus, since $B^{(n)}$ is a Noetherian S-module, so is $_S E$. It follows that E is a left Noetherian ring.

Next, we show that E is prime. Let $0 \neq f, g \in E$. The simplicity of B implies that the non-zero S-T-bimodule $Ef(B)$ of B must equal B. Thus $gEf \neq 0$, as required. Now we can improve on this to show that

(2) $\qquad\qquad\qquad\qquad E \text{ is a simple ring.}$

Observe that the result follows at once from (i) and (2). To prove (2), we first show that, writing C_S and C_E for the regular elements of S and E respectively,

(3) $\qquad\qquad\qquad\qquad\qquad C_S \subseteq C_E.$

Let $c \in C_S$ and $f \in E$. If $cf(B) = 0$, then $f(B)$ must be zero since $_S B$ is torsion-free, and so f is zero. Suppose that $fc = 0$. Then $f(cB) = 0$, so f induces a T-homomorphism f' from B/cB to B. But $u - \dim(cB_T) = u - \dim(B_T)$, so the image of f' is T-torsion. Therefore $f' = 0$, so $f = 0$ and (3) is proved.

Again let $c \in C_S$. By (3) $u - \dim(_S E) = u - \dim(_S Ec)$, and so $_S(E/Ec)$ is a torsion module. Therefore C_S is a left Ore set in E. Consider the ring $C_S^{-1} E$, a partial quotient ring of E in view of (3). Being finitely generated as a left $C_S^{-1}S$-module, $C_S^{-1}E$ is a left Artinian ring, and hence it's the full left quotient ring of the prime left Noetherian ring E. Thus, by Goldie's theorem [**32**, Theorem 2.3.6], [**18**, Theorem 5.12]

(4) $\qquad\qquad\qquad\qquad C_S^{-1} E \text{ is simple Artinian .}$

Now let I be a non-zero ideal of E. Then (4) shows that $I \cap S \neq 0$, so the simplicity of S forces $I = E$, and (2) is proved, as required.

Theorem 10 suggests the strategy of studying simple Noetherian rings by examining their Noetherian bimodules, with a view to finding numerical (and other) invariants of the rings. Given a simple Noetherian ring R one could, for example, study the behaviour of the class of Noetherian R-R-bimodules under the operations $\oplus$ and $\otimes_R$. Choosing among many open problems, we list only

QUESTION H. *Let R be simple Noetherian. What can be said—in general, and also for specific rings R—about the ring structure of the Grothendieck group of (projective) Noetherian R-R-bimodules?*

§3. Injective modules

Recall that the set $\mathrm{Ass}(M)$ of *associated primes* of an R-module M is the set of prime ideals P of R for which M has a P-prime submodule. Since every injective R-module is a direct sum of indecomposable injective R-modules [**18**, Corollary 4.20], the study of injective R-modules reduces at once to the case of an indecomposable one, E say. Thus E is uniform, and so has a unique associated prime P; there is therefore a cyclic P-prime submodule U of E with $E = E_R(U)$. In general, of course, U is not unique, but we do at least have a map $E \longrightarrow \mathrm{Ass}(E)$ from the isomorphism classes of indecomposable injective (right) R-modules to $\mathrm{Spec}(R)$. This map is surjective (since we can always choose U to be a uniform right ideal of R/P), but it is injective only if R is (right) fully bounded [**18**, Proposition 8.13].

To analyse the structure of E we shall need an analogue of the socle series in this setting. There are technical difficulties arising here which can most easily be finessed by imposing an extra hypothesis on the ring R.

HYPOTHESIS 11. *If P and Q are prime ideals of R with $Q \in r.cl(P)$ and $Q \subseteq P$, then $Q = P$.*

Thus Hypothesis 11 requires that Question F has a negative answer for R. If R satisfies the r.s.s.l.c. then Hypothesis 11 holds for R in view of Theorem 9. Hypothesis 11 also holds when R is the enveloping algebra of a finite dimensional Lie algebra, as explained in the remark following Question F.

Let us call a set X of prime ideals of R *incomparable* if $P \subseteq Q$ only if $P = Q$. We can thus restate Hypothesis 11 as: *Every clique of R is incomparable.*

DEFINITION AND LEMMA 12: (i) Let P be a prime ideal of R. Define subsets $X_n(P)$ of $r.cl(P)$ as follows: $X_1(P) = P$, and for $n \geq 1$, $X_{n+1}(P) = \{Q \in \operatorname{Spec}(R) \colon Q \rightsquigarrow\rightarrow I \in X_n(P)\}$.

(ii) Let M be an R-module for which $\cup\{r.cl(P) \colon P \in \operatorname{Ass}(M)\}$ is incomparable. The *first storey series* $\{F_n(M)\}$ and n^{th} *layer fundamental primes* $A_n(M)$ of M are defined as follows:

$$F_0(M) = 0; \quad A_1(M) = \operatorname{Ass}(M);$$

$F_1(M) = \{m \in M \colon \operatorname{Ann}(mR) \text{ contains an } \operatorname{Ass}(M)\text{-semiprime ideal }\}.$ For $n > 1$, set

$$A_n(M) = \operatorname{Ass}(M/F_{n-1}(M)) \cap \cup\{X_n(P) \colon P \in \operatorname{Ass}(M)\},$$

and put

$$F_n(M) = \{m \in M \colon \operatorname{Ann}(mR + F_{n-1}(M)/F_{n-1}(M)$$
$$\text{contains an } A_n(M)\text{-semiprime ideal }\}.$$

(iii) With M as in (ii), $F(M) = \cup_n F_n(M)$ is called the *first storey* of M. If $M = F(M)$ we say that M is *single-storied*.

(iv) The set $A(M) = \cup_{n \geq 1} A_n(M)$ is the set of *first storey primes* of M.

(v) The subset $A(P) = \cup_{n \geq 1} A_n(E(R/P))$ of $r.cl(P)$ is called the set of *first storey primes* of P; and we shall write $A_n(P)$ for $A_n(E(R/P))$, calling this the set of nth *layer primes* of P.

Using Lemma 4 it is not hard to see that $F_n(M)/F_{n-1}(M)$ is the unique biggest $A_n(M)$-semiprime submodule of $M/F_{n-1}(M)$. It is inconvenient but necessary to impose the incomparability condition on $\operatorname{Ass}(M)$ in Definition 12: Consider otherwise how one would define $F_1(M)$ for the **Z**-module $M = \mathbb{Q} \oplus C$ when C is a Prufer 2-group.

In understanding these definitions it helps to have in mind the commutative case, reviewed in Example 14 below. A useful tool here is the following lemma, which, simple though it is, lays bare the mechanism whereby the two-sided ideal structure of the ring influences the (one-sided) representation theory—for note that the R-module structure of the Hom-group below arises from the *left* action of R on J/I. The lemma is proved by applying $\operatorname{Hom}_R(-, E)$ to the exact sequence $0 \longrightarrow J/I \longrightarrow R/I \longrightarrow R/J \longrightarrow 0$. Note that the submodule $\operatorname{Ann}_E(P)$ of E appearing in the lemma is an injective R/P-module.

LEMMA 13. (Rosenberg and Zelinsky, [37]). *Let E be an injective R-module with $\text{Ass}(E) = \{P\}$, and let I and J be ideals of R with $JP \subseteq I \subseteq J \subseteq P$. Then*

$$\text{Ann}_E(I)/\text{Ann}_E(J) \cong \text{Hom}_R(J/I, \text{Ann}_E(P)).$$

EXAMPLE 14. (Matlis, 1958, [31]). *Let P be a non-zero prime ideal of the commutative Noetherian domain R, and let $M = E_R(R/P)$.*

As we have already noted, $cl(P) = \{P\}$. Since M is a faithful R-module $A_n(M) = \{P\}$ for all $n \geq 1$. As P satisfies the AR property, (or simply by quoting Lemma 4), every finitely generated submodule of M is annihilated by a power of P [18, Lemma 11.11], so that $M = F(M)$ and $F_n(M) = \text{Ann}_M(P^n)$ for each $n \geq 1$. Moreover $\text{Ann}_M(P) = F_1(M) = E_{R/P}(R/P)$ is the field of fractions K of R/P. Thus Lemma 13 implies that for each $n > 1$, $F_n(M)/F_{n-1}(M)$ is a finite dimensional K-vector space; and indeed the dimensions of these spaces can be read off from the Poincaré series of the local ring R_P.

Returning now to the notation of Definition 12, we examine how Example 14 generalises to an arbitrary ring with the s.s.l.c. Recall that Theorem 9 guarantees that Hypothesis 11 is in force here. Thus the definition applies to an R-module M with $\text{Ass}(M) = \{P\}$, and since Hypothesis 11 excludes the possibility that there exist primes I and Q in $r.cl(P)$ with $I \angle Q$, one sees from Lemma 4 that $F(M)$ is built up from $\text{Ann}_M(P)$ by a sequence of extensions of type (a); indeed $F(M)$ is defined so as to contain *all* submodules of M obtainable in this way. This therefore gives (i) of the next result.

THEOREM 15. (Brown, Warfield [13]). *Let R be a Noetherian ring with the s.s.l.c. Let P be a prime ideal of R, let U be a uniform right ideal of R/P, and set $E = E_R(U)$.*

(i) *E is single-storied; that is, $E = F(E) = \cup_n F_n(E)$.*
(ii) *$F_1(E)$ is the irreducible $Q(R/P)$-module.*
(iii) *For $n > 1$, $F_n(E)/F_{n-1}(E)$ is a direct sum of non-zero $Q(R/I)$-modules, as I ranges through $A_n(P)$.*

Clearly $A_1(P) = \{P\}$, and Lemma 4(ii), the converse to the Main lemma, shows that $A_2(P) = X_2(P)$, but in general $A_n(P) \subsetneq X_n(P)$ for $n > 2$. We shall discuss further in Theorem 26 the problem of recognising the first storey primes of P.

At first glance the picture given by Theorem 15 looks very similar to the commutative case discussed in Example 14, with only the "twists" supplied by the second layer links complicating matters, but in fact for a noncommutative ring $E(R/P)$ can be much "fatter" than in Example 14, in two ways: First, the sets $A_n(P)$ can be infinite [**39**, 4.4], [**25**, Ex.6.2.18], though as subsets of $r.cl(P)$ they are necessarily countable. Second, there may exist $n > 1$ and $I \in A_n(P)$ such that the multiplicity of the irreducible $Q(R/I)$-module in $F_n(E(R/P))/F_{n-1}(E(R/P))$ is infinite; indeed this already happens for R Artinian and $n = 2$ by results of Rosenberg, Zelinsky and Cohn [**37**], [**15**]. But there is evidence to support the belief that this sort of behaviour is fairly unusual; for more details see [**13**, 5.11 and 5.12].

A further complication in the noncommutative case is caused by the fact that, when R is not FBN, Theorem 15 does not begin to cover all the indecomposable injective R-modules. Surprisingly, though, at least so far as "twisting" is concerned, when R has the s.s.l.c. Theorem 15 tells the whole story—this is the content of Theorem 16, the main result of this section, which can be summed up as: "Primes are persistent".

THEOREM 16. (Brown, Warfield [**13**]). *Let R be a Noetherian ring with the s.s.l.c. Let P be a prime ideal of R, let U' be any uniform P-prime R-module, and set $E' = E_R(U')$.*

(i) *E' is single-storied; that is, $E' = F(E') = \cup_{n \geq 1} F_n(E')$.*

(ii) *$F_1(E')$ is the R/P-injective hull of U'.*

(iii) *For each $n > 1$, $F_n(E')/F_{n-1}(E')$ is the direct sum of non-zero R/I-modules, as I ranges through $A_n(P)$.*

Here, (ii) is immediate from the definitions, and (i) is in essence a restatement—via Lemma 4—of part of the hypothesis that R has the s.s.l.c. The key point is contained in (iii): The associated primes of $E'/F_{n-1}(E')$ are precisely the associated primes of $E/F_{n-1}(E)$; or, in brief, $A_n(E') = A_n(P)$. In fact, let's assume for simplicity that the subsets $A_n(P)$ of $\text{Spec}(R)$ are all finite, and define, for $n \geq 1$, $S_n(P) = \cap\{I : I \in A_n(P)\}$, $T_n(P) = S_n(P)S_{n-1}(P)\ldots S_1(P)$ and $U_{n+1}(P) = S_{n+1}(P)S_n(P)\ldots S_2(P)$. Now fix $n \geq 1$ and $I \in A_{n+1}(P)$, and define the ideal

$$T(I, n, P) = \{r \in T_n(P) \cap U_{n+1}(P) : Ir \subseteq T_{n+1}(P)\}.$$

Then the proof of Theorem 16(iii) proceeds by using Lemma 13 to show that

(5) $$\text{Hom}_R(T(I, n, P)/T_{n+1}(P), F_1(E'))$$

is isomorphic to the I-homogeneous component of $F_{n+1}(E')/F_n(E')$. (A similar formulation can be obtained even when the sets $A_n(E)$ are infinite.)

Thus, associated with each prime ideal P of R is a "skeleton", each bone of which is labelled by a bimodule of ideals with right annihilator P, of the form $T(I, n, P)/T_{n+1}(P)$ defined above. Given a fully faithful R/P-module U', we construct its injective hull E' by first forming its R/P-hull $F_1(E')$, and then inserting a non-zero subfactor into E' for each bone of the skeleton according to the formula (5).

Note one rather obvious consequence of Theorem 15(iii): When U is a torsion-free R/P-module the I-homogeneous components of the layers $F_n(E)/F_{n-1}(E)$ of its injective hull are R/I-injective, for every prime ideal I in $A_n(P)$. Presumably this does not extend to the torsion modules which are the subject of Theorem 16, but at present no counterexample is known; this prompts the next question.

QUESTION I. *(In the notation of Theorem 16), are the I-homogeneous components of the layers $F_n(E')/F_{n-1}(E')$ injective R/I-modules?*

Here are some cases where a counterexample should not be sought. (A Noetherian ring R is *polynormal* if for all ideals $I \subsetneq J$ of R there exists $x \in J \setminus I$ with $I + xR = I + Rx$.)

THEOREM 17. *Assume the hypotheses and notation of Theorem 16. If P is a maximal ideal, or if R is a polynormal ring, then the I-homogeneous components of the layers $F_n(E')/F_{n-1}(E')$ are injective R/I-modules.*

PROOF: Suppose that P is a maximal ideal of R. Fix $I \in A_{n+1}(P)$. By (5), the I-homogeneous component we are interested in is

$$H := \operatorname{Hom}_R(T(I, n, P)/T_{n+1}(P), F_1(E')).$$

Let A be a right R/I-module. By [**14**, VI.5.1], (noting that $F_1(E')$ is R/P-injective),

$$\operatorname{Ext}^1_{R/I}(A, H) \cong \operatorname{Hom}_{R/P}(\operatorname{Tor}^{R/I}_1(A, T(I, n, P)/T_{n+1}(P)), F_1(E')).$$

Theorem 10 ensures that the R/I-R/P-bimodule $T(I, n, P)/T_{n+1}(P)$ is (left) R/I-projective, so that the left side of the above isomorphism is zero. Thus H is R/I-injective.

The proof is similar when R is polynormal [**13**, 6.2].

§4. Indecomposable injectives without the second layer condition

With the results of §§1 and 3 the broad outlines of the representation theory of Noetherian rings with the s.s.l.c. begins to take shape—enough is known, at least, to formulate a coherent series of open questions. But this is not the case for rings without the s.l.c. where the open questions on bimodules from §2 play a debilitating role. We avoid these problems by imposing a condition stronger than Hypothesis 11 throughout this section:

HYPOTHESIS 18. *The Noetherian ring R possesses a dimension function defined on the isomorphism classes of non-zero finitely generated right and left R-modules, which takes non-negative integer values and is exact, symmetric and prime separating (as defined following Question F).*

The Gel'fand-Kirillov dimension, for example, satisfies these conditions when R is the enveloping algebra of a finite dimensional Lie algebra [**26**]. If Question D has a positive answer then the Krull dimension will do in Hypothesis 18, for every Noetherian ring R with finite Krull dimension. Naturally, we ask:

QUESTION J. *Are the results of §4 true in the absence of Hypothesis 18?*

The chief obstruction to answering this question is the lack of a workable partial ordering on cliques, bearing in mind that Theorem 9 does not extend to the present context. For how else can we generalise the following definition?

DEFINITION 19: Assume Hypothesis 18. Let M be an R-module. Let $t_1 = \min\{d(R/P): P \in Ass(M)\}$. Set $S_0(M) = 0$, and the *first storey* of M,

$$S_1(M) = \{m \in M : d(R/\operatorname{Ann}(mR)) = t_1\}.$$

For $i > 1$, (having defined $S_{i-1}(M)$), define the ith *storey* of M, $S_i(M)/S_{i-1}(M)$, to be the first storey of $M/S_{i-1}(M)$. Set $t_i = d(R/Q)$, where $Q \in \operatorname{Ass}(S_i(M)/S_{i-1}(M))$, and call t_i the *annihilator codimension of the ith storey*.

Thus the idea is to begin building the ith storey by adding to $S_{i-1}(M)$ all the modules annihilated by primes of minimal d-codimension amongst the annihilator prime ideals of $M/S_{i-1}(M)$, to get, say, $U/S_{i-1}(M)$. Then complete the ith storey by including all the essential extensions of submodules of $U/S_{i-1}(M)$ formed by a sequence of "second layer" extensions as in

Lemma 4(a). This is made precise in (iv) of the next lemma, showing that the terminology "first storey" of Definitions 12 and 19 is consistent.

The following facts are easy consequences of the definition, the properties of d, and of Lemma 4. Their proofs can be found in [**29**].

LEMMA 20. *(i) Each $S_i(M)$ is a submodule of M.*

(ii) $t_1 < t_2 < \cdots \leq d(R)$. Thus there exists n (with $n \leq d(R)$) such that $S_n(M) = M$.

(iii) Let N be a submodule of M, and let $i \geq 0$. Then $S_i(N) = N \cap S_i(M)$.

(iv) Let $i \geq 1$, and suppose that $M/S_{i-1}(M)$ satisfies the incomparability hypothesis imposed on M in Definition 12(ii)-(iv). Then $S_i(M)/S_{i-1}(M)$ is precisely $F(M/S_{i-1}(M))$, (as defined in Definition 12(iii)).

Following Definition 12 we can refine each storey of an R-module M over a ring satisfying Hypothesis 18. Here are the details:

NOTATION 21: (i) Let M and R be as in Definition 19. For $i \geq 1$ and $j \geq 0$ denote the jth term of the first storey series of $S_i(M)/S_{i-1}(M)$ by $F_{ij}(M)$. Thus, for each $i \geq 1$,

$$S_{i-1} = F_{i0}(M) \subseteq F_{i1}(M) \subseteq \cdots \subseteq F_{i(j-1)}(M) \subseteq F_{ij}(M)$$
$$\subseteq \cdots \subseteq \cup_{j \geq 0} F_{ij}(M) = S_i(M).$$

(ii) For $j \geq 1$ denote the set of associated primes of $F_{ij}(M)/F_{i(j-1)}(M)$ by $A_{ij}(M)$. For each $i \geq 1$ denote $\cup_{j \geq 1} A_{ij}(M)$ by $\mathrm{Fund}_i(M)$, the ith *storey prime ideals of M*. $\mathrm{Fund}(M) = \cup_{i \geq 1} \mathrm{Fund}_i(M)$ is called the set of *fundamental primes of M*.

(iii) Let P be a prime ideal of R and set $E = E_R(R/P)$. We write $A_{ij}(P)$, $\mathrm{Fund}_i(P)$ and $\mathrm{Fund}(P)$ instead of $A_{ij}(E)$, $\mathrm{Fund}_i(E)$ and $\mathrm{Fund}(E)$ respectively; $\mathrm{Fund}_i(P)$ is called the set of ith *storey prime ideals of P*, and $\mathrm{Fund}(P)$ is called the set of *fundamental prime ideals of P*.

Note that P has the s.l.c. if and only if $\mathrm{Fund}(P) = \mathrm{Fund}_1(P)$ if and only if $E = E_R(R/P)$ is single-storied if and only if the fundamental primes of P are just the first storey primes of P, (since $\mathrm{Ass}(E/S_i(E))$ is disjoint from $\cup_{j \leq i} \mathrm{Fund}_j(P)$ by Hyposthesis 18 and Lemma 4(b)). Recall from Theorem 16(i) that if R has the s.s.l.c. then all indecomposable injective R-modules are single-storied. In view of Lemma 20(iii) the same is true of all uniform R-modules; and in fact, (over a ring which satisfies the s.s.l.c. and Hypothesis 18), the stories of an arbitrary module serve simply to filter the module according to the d-codimensions of its annihilator primes. Surprisingly, the same is true without the s.s.l.c. To state the result, we need

DEFINITION 22: *Let M be a module over the Noetherian ring R. The annihilator spectrum of M,*

$$\text{Ann-spec}(M) = \{Q \in \text{Spec}(R) \colon Q = \text{Ann}(N), \, 0 \neq N \subseteq M,$$

$$N \text{ finitely generated } \}.$$

Lemma 4 shows that if R has the s.s.l.c. then $\text{Ann-spec}(M)$ is just the set $\text{Ass}(M)$ of associated primes of M defined at the start of §3; whereas (b) of the same result makes clear that, without s.s.l.c., $\text{Ass}(M) \subsetneq \text{Ann-spec}(M)$ in general.

Now let R satisfy Hypothesis 18 and let M be an R-module. The first question to address in trying to understand the storeys of M is: For each i, what is $\text{Ass}(S_i(M)/S_{i-1}(M))$? A prime Q in this set corresponds to a fully faithful cyclic R/Q-submodule of $S_i(M)/S_{i-1}(M)$, say $mR + S_{i-1}(M)/S_{i-1}(M)$. Standard methods (using Lemma 4) applied to the submodule mR of M show that Q must be jump linked to a prime $I \in \text{Fund}(S_{i-1}(M)) = \cup_{t \leq i-1} \text{Fund}_t(M)$. Therefore Lemma 4(b) implies that Q is an annihilator prime of M/X, for some $X \subsetneq S_{i-1}(M)$. But in fact X here can always be taken to be zero. This is the content of

THEOREM 23. (Lenagan, Warfield, [29]). *Let R satisfy Hypothesis 18, and let M be an R-module. Then*

$$\text{Ann-spec}(M) = \cup_{i \geq 1} \text{Ass}(S_i(M)/S_{i-1}(M)).$$

Using Notation 21 we can re-write this as

$$\text{Ann-spec}(M) = \cup_{i \geq 1} A_{i1}(M).$$

This is really a result about finitely generated modules; to prove it one passes by standard reductions to a set-up which is not much more complex than the following: M is a uniform R-module with submodules A and B such that A, B/A and M/B are irreducible with annihilators P, I and Q respectively. Suppose that $I \rightsquigarrow\rightarrow P$ (with corresponding module B), and $Q \angle I$ (with corresponding module M/A). Thus $B(IP) = 0$ and $(M/A)Q = 0$ (so that $MQ \subseteq A$). Thus here we have $S_1(M) = B$ and $S_2(M) = M$; $\text{Ass}(M) = \{P\}$, $\text{Fund}_1(M) = \{P, I\}$, and $\text{Fund}_2(M) = \{Q\} = \text{Ass}(S_2(M)/S_1(M))$. What has to be proved is that $MQ = 0$. This is achieved by arguments like those deployed to prove Lemma 4.

Here is the most familiar context where some of these ideas come into play. (We refer to [16] for the necessary properties of enveloping algebras.)

EXAMPLE 24: Let $R = U(\mathbf{sl}(2,\mathbb{C}))$, P its augmentation ideal, and $E = E_R(R/P)$, the injective hull of the trivial module.

(a) Let Ω be the Casimir element of R, (a non-zero element of the centre of R), and set $Q = \Omega R$, so Q is a primitive ideal contained in P. In fact, P is the only proper ideal of R strictly containing Q. Let $I \sim\!\sim\!\longrightarrow P$. Then $Q \subseteq I$ (since Ω is central), and R/I must have finite $\mathbb{C}$-dimension; therefore $I = P$ is the only possibility. But $P = P^2$, so $r.cl(P) = \{P\}$.

(b) By Lemma 8, $Q \angle P$. In fact, an essential extension of R/P affording this jump link is afforded by the dual (in the category O) of a Verma module having $R/P = \mathbb{C}$ as its unique irreducible image. The centrality of Ω ensures that $\{0\}$ is not jump linked to P; and since Q is the only non-zero prime ideal strictly contained in P, it must be the only ideal jump linked to P.

(c) Since Q is centrally generated, $r.cl(Q) = \{Q\}$, and no prime is jump linked to Q. Graphically, we could represent our conclusions from (a), (b) and (c) by:

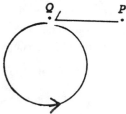

(d) *The structure of E.* Since $r.cl(P) = \{P\}$,

$$S_1(E) = E_{R/P}(R/P) = R/P = \mathbb{C}.$$

By (b), $\mathrm{Ass}(E/S_1(E)) = Q$, so

$$F_{21}(E) = \{e \in E \colon eQ \subseteq S_1(E)\}.$$

By Lemma 4(b), $F_{21}(E)Q = 0$, and so

(6) $$F_{21}(E) = E_{R/Q}(R/P).$$

Since E is Ω-divisible, we calculate, noting (c), that

(7) $$F_{2i}(E)/F_{2(i-1)}(E) \cong F_{21}(E)/(R/P)$$

for all $i \geq 1$, the isomorphism being given by multiplication by Ω^{i-1}. And by (b) and (c),

(8) $$E = S_2(E) = \cup_{i \geq 0} F_{2i}(E).$$

There is thus a chain

$$(9) \quad 0 \subset R/P = S_1(E) = F_{20}(E) \subset F_{21}(E) \subset \ldots$$
$$\subset \cup_{i \geq 0} F_{2i}(E) = S_2(E) = E,$$

where all the factors after the first are isomorphic.

(e) By choosing different Cartan subalgebras in $\mathbf{sl}(2, \mathbf{C})$ we can find infinitely many distinct Verma modules whose duals have socle R/P. It follows from this that $F_{21}(E)/(R/P)$ has infinite uniform dimension. The isomorphism (7) extends this observation to all higher factors in the chain (9).

(f) From (6), (7) and (8) we can state

$$\text{Ann-spec}(E) = \{P, Q\} \supsetneq \{P\} = \text{Ass}(E).$$

It would be interesting to study other enveloping algebras in the above style. Presumably indecomposable injective modules with any prescribed finite number of annihilator primes could be constructed in this way; but it may be harder to settle the following:

QUESTION K. *Can $|\text{Ann-spec}(E)| = \infty$, for an indecomposable injective E over a Noetherian ring? Is Ann-spec(E) always countable?*[2]

Even when Hypothesis 18 holds Question K is open, because although in this case modules can only have finitely many stories, it is not known whether $S_i(E)/S_{i-1}(E)$ can have infinitely many associated primes, for some $i > 1$. Notwithstanding this and other gaps in our understanding, there is a beautiful generalisation of the Prime Persistence Theorem 16 to the present setting:

THEOREM 25. (Lenagan, Warfield [29]). *Assume Hypothesis 18, and let E' be an indecomposable injective R-module with $\text{Ass}(E') = P$.*

(i) $\text{Fund}_1(E') = \{Q \in \text{Spec}(R) \colon Q \approx P \text{ via } B/A, \text{ with } d(R/A) = d(R/P)\}.$

    Thus (as in Theorem 16), $\text{Fund}_1(E')$ depends only on P, not on E'.

(ii) *For $i \geq 1$,*

$$\text{Fund}_i(E') = \cup\{\text{Fund}_1(Q) \colon Q \in \text{Ass}(S_i(E')/S_{i-1}(E'))\}.$$

(iii) $\text{Fund}(E') = \cup\{\text{Fund}_1(Q) \colon Q \in \text{Ann-spec}(E')\}.$

---

[2]See note added in proof, page 24.

In fact (i) can be improved to give a full generalisation of Theorem 16 (under Hypothesis 18), the missing information on the layers (namely the make-up of the sets $A_{1j}(P)$ for $j \geq 1$) being implicit in the proof in [**29**]. And (ii) can be improved to read: *For $i \geq 1$ and $j \geq 0$,*

$$A_{ij}(E') = \cup\{A_{1j}(Q) \colon Q \in \text{Ass}(S_i(E')/S_{i-1}(E'))\}.$$

As usual, the proofs of (i) and (ii) are local arguments in the style of Lemma 4, making use of variants of the Rosenberg-Zelinsky Lemma 13 to show the right-hand sets are in the left side.

Part (i) is related to the following result which pins down $Fund(P)$ within $r.cl(P)$ for a ring with the s.l.c.

THEOREM 26. (Lenagan, Letzter [**28**]). *Let R be a Noetherian ring with s.l.c. and let P be a prime ideal of R. Then $Q \in \text{Fund}_1(P) = \text{Fund}(P)$ if and only if $Q \approx P$.*

The reader will doubtless be tempted to formulate conjectures combining generalisations of Theorems 25 and 26 which would be valid for *any* Noetherian ring. When doing so, note that the extra condition on codimensions in Theorem 25(i) is certainly needed, as is illustrated by an example of Stafford [**39**, Theorem 4.1] of a Noetherian ring R with primes Q and P such that $Q \approx P$ but $Q \notin r.cl.(P)$, so that $Q \notin \text{Fund}_1(P)$. Necessarily, in view of Theorem 26, this ring does not satisfy s.l.c. The point is that in a ring with s.l.c. every ideal link can be raised to "just below" the prime ideals involved (essentially because such rings possess a version of primary decomposition), but this is not possible in every Noetherian ring.

Notice that Theorem 25 does not guarantee that under its hypotheses $\text{Fund}(E) = \text{Fund}(E')$ for indecomposable injective modules with the same assassinator. For this to be true we need a positive answer to

QUESTION L. *Let E and E' be indecomposable injective modules with the same associated prime over a Noetherian ring R. Is $\text{Ann-spec}(E) = \text{Ann-spec}(E')$?*

Perhaps the correct approach to Question L is via

QUESTION M. *Is there a way to recognise "ring theoretically" when $Q \angle P$?*

Theorem 25(ii) hints that indecomposable injective modules may be built by stacking various first storeys on top of one another, to form a tower. We formulate this as

QUESTION N. *Assume the hypotheses and notation of Theorem 25. For $i > 1$, is $S_i(E)/S_{i-1}(E)$ isomorphic to the first storey of some module?*

§5. Rings satisfying a polynomial identity

Suppose that R is a ring satisfying the s.l.c., and let P be a prime ideal of R. We know from Lemma 4 that the set Fund(P) of primes occurring as the associated primes of the layers $F_i(E)/F_{i-1}(E)$ of $E = E_R(R/P)$ is contained in $r.cl(P)$. In fact Theorem 26 shows that Fund(P) is exactly the set of primes Q with $Q \approx P$. How does Fund(P) compare with $r.cl(P)$? At one extreme we can have Fund$(P) = r.cl.(P)$, as for instance when $r.cl(P) = \{P\}$, and—in certain cases at least—when R is the enveloping algebra of a solvable Lie algebra. Rings satisfying a polynomial identity lie at the other extreme:

THEOREM 27. (Mueller, [**34**, Theorem 7]). *If R is a Noetherian PI-ring and P is a prime ideal of R, then $|\operatorname{Fund}(P)| < \infty$.*

The proof of Theorem 27 proceeds by reducing first to the case where R is semiprime, by invoking Noetherian induction and applying the Rosenberg-Zelinsky exact sequence of Lemma 13. (But this is not entirely routine — a trick is needed involving a switch from right to left modules; see [**34**].) A second reduction — and this one really is a routine use of Lemma 13 — will take us from the semiprime to the prime case.

For R prime (or even semiprime), one can, following Mueller, use the ubiquity of non-zero central elements to complete the proof by Noetherian induction. We offer instead the following approach which has the merits of globally bounding $|\operatorname{Fund}(P)|$ and of suggesting directions for future work. It employs the *trace ring* $T(R)$ of the prime Noetherian PI-ring R, the subring of the full quotient ring Q of R obtained by adjoining to R the coefficients (from the centre C of Q) of the characteristic polynomials of the elements of R, when the latter are viewed as C-linear transformations of Q; details can be found in [**32**, §13.9], for example.

THEOREM 28. (Lenagan, Letzter [**28**]). *Let P be a prime ideal of the prime Noetherian PI-ring R. Suppose that the trace ring $T(R)$ of R can be generated by m elements which centralise R, and that the PI-degree of R is n. Then*

$$|\operatorname{Fund}(P)| \leq mn.$$

In the sketch of the proof of this theorem which follows, we retain the hypotheses and notation of Theorem 28 and the sentence preceding it. Since

$$(10) \qquad T(R) = \sum_{i=1}^{m} c_i R$$

for elements $c_1 \ldots c_m$ of $C \cap T(R)$ [32, Proposition 13.9.11(i)], $T(R)$ is a Noetherian ring whose representation theory is closely connected to that of R; and since $T(R)$ is integral over its centre Z [32, Proposition 13.9.5], we may hope to find $T(R)$ easier to study than R. Theorem 28 is a consequence of the following three facts, of which (ii) illustrates the second of these points, and (i) and (iii) the first.

(i) [32, Theorem 10.2.7(i)], [36, Theorem 3.4] *Let P be a prime ideal of R. There are at most m, (and there is at least one) prime ideal(s) P_i' of $T(R)$ minimal over $PT(R)$; and they all satisfy $P_i' \cap R = P$.*

The ideals P_i' in (i) are said to *lie over P.*

(ii) [6, Proposition 3], [4, Proposition 5], [5, Lemma 2] *Let P' be a prime ideal of $T(R)$. Then*

$$cl(P') = \{Q' \in \mathrm{Spec}(T(R)) : Q' \cap Z = P' \cap Z\}.$$

Consequently,

$$|cl(P')| \leq n.$$

The cord connecting the representation theory of R with that of $T(R)$ is provided by Theorem 26 and

(iii) [28, Theorems 3.1, 3.2] *Let P and Q be primes of R. Then $Q \approx P$ if and only if there are primes Q' and P' of $T(R)$ lying over Q and P respectively, with $Q' \approx P'$.*

Here are the bones of a proof of (iii), (different from the one in [28]). Suppose first that $Q' \approx P'$, where these are primes of $T(R)$ lying over Q and P, via the ideal link B'/A', say. It is clear from (10) that there is a non-zero central element d of R with $dT(R) \subseteq R$. Thus dA' and dB' are ideals of R with $dB'/dA' \cong B'/A'$ as R-modules, so affording an ideal link between $Q' \cap R = Q$ and $P' \cap R = P$.

Conversely, suppose that $Q \approx P$, so that by Theorem 26 Q occurs as a fundamental prime of $E = E_R(R/P)$. Thus there exist finitely generated R-submodules $X \subseteq Y$ in E with $\mathrm{Ann}_R(Y/X) = Q$. Apply the exact functor $\mathrm{Hom}_R(-, E)$ to the sequence $0 \longrightarrow R \overset{i}{\longrightarrow} T(R) \longrightarrow T(R)/R \longrightarrow 0$ of right R-modules, yielding

$$0 \longrightarrow \mathrm{Hom}_R(T(R)/R, E) \longrightarrow \mathrm{Hom}_R(T(R), E) \overset{i*}{\longrightarrow} E \longrightarrow 0.$$

Set $E' = \mathrm{Hom}_R(T(R), E)$, an injective $T(R)$-module by [**14**, VI.5.1]. It is easy to see that $\mathrm{Ass}(E')$ consists of primes of $T(R)$ lying over P, and that $i^{*-1}(Y)/i^{*-1}(X)$ generates a $T(R)$-subfactor of E' with an assassinator prime Q' lying over Q. That is, $Q' \in \mathrm{Fund}(E')$, and a final application of Theorem 26 completes the proof.

Theorem 28 follows at once from (i), (ii) and (iii).

There is obviously scope for further work here. Most glaringly, it appears to be unknown at present whether $cl(P')$ and $\mathrm{Fund}(P')$ coincide. Thus we ask:

QUESTION O. *Let T be a prime Noetherian PI-ring, integral over its local centre Z, and let P' and Q' be maximal ideals of T. Is $Q' \approx P'$?*

Note added in proof: Following circulation of this paper, J.T. Stafford made progress on two of the listed questions. (1) There exists a Noetherian domain with a clique X which does not satisfy the second layer condition, but the intersection of the members of X is 0; this answers Question C in the negative. (2) There exists a Noetherian domain R sith an ideal P such that R/P is a division ring, and distinct primitive ideals $K_i(i \geq 1)$ contained in P, such that $E_R(R/P)$ contains a faithful R/K_i-module of composition length two, for each i. Thus (2) gives a positive answer to the first part of Question K; the second part remains open. Details of Stafford's constructions will appear elsewhere.

24   K.A. Brown

REFERENCES

1. A. D. Bell, *Localization and ideal theory in iterated differential operator rings*, J. Algebra **106** (1987), 376–402.
2. A.D. Bell, *Localization and ideal theory in Noetherian strongly group-graded rings*, J. Algebra **105** (1987), 76–115.
3. A.D. Bell, *Notes on localization in Noetherian rings*, Cuadernos de Algebra **9** (1988), Universidad de Granada.
4. A. Braun, *An additivity principle for PI-rings*, J. Algebra **96** (1985), 433–441.
5. A. Braun and L.W. Small, *Localization in prime Noetherian PI-rings*, Math. Zeit. **193** (1986), 323-330.
6. A. Braun and R.B. Warfield, Jr, *Symmetry and localization in Noetherian prime PI rings*, J. Algebra **118** (1988), 322–335.
7. K.A. Brown, *Module extensions over Noetherian rings*, in "Proc. Conf. on Noetherian Rings and Rings with Polynomial Identities (Durham University)," Mimeo., University of Leeds, 1979, pp. 142–143.
8. K.A. Brown, *Module extensions over Noetherian rings*, J. Algebra **69** (1981), 247–260.
9. K.A. Brown, *The structure of modules over polycyclic groups*, Proc. Cambridge Phil. Soc. (to appear).
10. K.A. Brown, *Ore sets in enveloping algebras*, Comp. Math. **53** (1984), 347–367.
11. K. A. Brown, *On the representation theory of solvable Lie algebras II: The abelian group attached to a prime ideal*, J. London Math. Soc. (to appear).
12. K.A. Brown and F. du Cloux, *On the representation theory of solvable Lie algebras*, Proc. London Math. Soc **57** (1988), 284–300.
13. K.A. Brown and R.B. Warfield, Jr, *The influence of ideal structure on representation theory*, J. Algebra **116** (1988), 294–315.
14. H. Cartan and S. Eilenberg, "Homological Algebra," Princeton University Press, 1956.
15. P. Cohn, *Quadratic extensions of skew fields*, Poc. London Math. Soc. **11** (1961), 531–566.
16. J. Dixmier, "Algebres Enveloppantes," Cahiers Scientifiques 37,Gauthier-Villars, Paris, 1974.
17. K.R. Goodearl and A.H. Schofield, *Non-Artinian essential extensions of simple modules*, Proc. Amer. Math. Soc. **97** (1986), 233–236.
18. K.R. Goodearl and R.B. Warfield, Jr, "An Introduction to Noncommutative Noetherian Rings," London Math. Soc. Student Texts No. 16, Cambridge University Press, Cambridge, 1989.
19. A.V. Jategaonkar, *A counterexample in ring theory and homological algebra*, J. Algebra **12** (1969), 418–440.
20. A. V. Jategaonkar, *Injective modules and classical localization in Noetherian rings*, Bull. Amer. Math. Soc. **79** (1973), 152–157.
21. A.V.Jategaonkar, *Injective modules and localization in noncommutative Noetherian rings*, J. Algebra **30** (1974), 103–121.
22. A.V. Jategaonkar, *Noetherian bimodules*, in "Proc. Conf on Noetherian Rings and Rings with Polynomial Identities (Durham University)," Mimeo., University of Leeds, 1979, pp. 158–169.

23. A.V. Jategaonkar, *Noetherian bimodules, primary decomposition, and Jacobson's conjecture*, J. Algebra **71** (1981), 379–400.
24. A.V. Jategaonkar, *Solvable Lie algebras, polycyclic-by-finite groups, and bimodule Krull dimension*, Comm. in Algebra **10** (1982), 19–69.
25. A.V. Jategaonkar, "Localization in Noetherian Rings," London Math. Soc. Lecture Note Series No. 98, Cambridge University Press ,Cambridge, 1986.
26. G. Krause and T.H. Lenagan, "Growth of Algebras and Gelfand-Kirillov Dimension," Research Notes in Math., No. 116, Pitman, London, 1985.
27. T.H. Lenagan, *Artinian ideals in Noetherian rings*, Proc. Amer. Math. Soc. **51** (1975), 499–500.
28. T.H. Lenagan and E.S. Letzter, *The fundamental prime ideals of a Noetherian prime PI ring*, Proc. Edin. Math. Soc. **33** (1990), 113–121.
29. T.H. Lenagan and R.B. Warfield, Jr, *Affiliated series and extensions of modules*, (to appear).
30. E. S. Letzter, *Prime ideals in finite extensions of Noetherian rings*, J. Algebra **135** (1990), 412–439.
31. E. Matlis, *Injective modules over Noetherian rings*, Pac. J. Math. **8** (1958), 511–528.
32. J.C. McConnell and J.C. Robson, "Noncommutative Noetherian Rings," Wiley-Interscience, New York, 1987.
33. B.J. Mueller, *Localization in fully bounded Noetherian rings*, Pac. J. Math. **67** (1976), 233–245.
34. B.J. Mueller, *Two-sided localization in Noetherian PI rings*, J. Algebra **63** (1980), 359–373.
35. D.G. Poole, *Localization in Ore extensions of commutative Noetherian rings*, J. Algebra **128** (1990), 434–445.
36. J.C. Robson, *Prime ideals in intermediate extensions*, Proc. London Math. Soc. **44** (1982), 372–384.
37. A. Rosenberg and D. Zelinsky, *Finiteness of the injective hull*, Math. Zeit. **70** (1958), 372–380.
38. G. Sigurdsson, *Links between prime ideals in differential operator rings*, J. Algebra **102** (1986), 260–283.
39. J.T. Stafford, *On the ideals of a Noetherian ring*, Trans. Amer. Math. Soc. **289** (1985), 381–392.
40. J.T. Stafford, *The Goldie rank of a module*, in "Noetherian Rings and Their Applications," (edited by L.W. Small) Math. Surveys and Monographs No. 24, American Mathematical Society, Providence, RI., 1987, pp. 1–20 1987.

Mathematics Dept.
University of Glasgow,
University Gardens, Glasgow G12 8QW
United Kingdom

Some Ring Theoretic Techniques and Open Problems in Enveloping Algebras

ANTHONY JOSEPH

§1. Introduction

The theory of enveloping algebras of Lie algebras underwent an explosive development during the period 1975-1985. This was due in part to the very rich structure afforded by the semisimple case and in part to the ever growing range of available techniques. This resulted in a certain suffocation and the subject is only now emerging from a quiescent phase. The resurgence is also due to interest in the related fields of graded Lie algebras, Kac-Moody algebras and Quantum groups.

The aim of these lectures is to review some of the simplest arguments which were successful. It is a happy accident that these mainly came from ring theory. Our emphasis will be on describing new or very recent results which can be so obtained. This will inevitably cause some distortion in our presentation; but this is certainly preferable to regurgitating well-known facts. Attention will be drawn to some important open problems.

Unless otherwise specified the base field k will be assumed to be of characteristic zero and with the particular exception of section 8 all Lie algebras will be assumed finite dimensional. We shall adopt the essentially standard notation of [14]. In particular for each Lie algebra $\mathfrak{a}$ we denote by $U(\mathfrak{a})$ its enveloping algebra, by ad $a : a \in \mathfrak{a}$ the derivation $x \mapsto ax - xa$ of $U(\mathfrak{a})$, by gr the gradation pertaining to the canonical filtration of $U(\mathfrak{a})$. Then $\mathrm{gr}(U(\mathfrak{a}))$ identifies with the symmetric algebra $S(\mathfrak{a})$ over $\mathfrak{a}$ which in turn identifies with the algebra of polynomial functions on the dual $\mathfrak{a}^*$.

§2. Local finiteness

2.1. It is a basic fact that ad $a : a \in \mathfrak{a}$ is a locally finite derivation of $U(\mathfrak{a})$. Now suppose we have a ring A with a locally nilpotent derivation ∂. Given $a \in A$ such that $\partial a = 1$ one easily checks that ad $a : x \mapsto ax - xa$ restricts to a derivation of the invariant ring A^∂. We may then form the skew polynomial extension $A^\partial \# k[a]$. An easy induction argument based on the identity $\partial^n a^n = n!$ gives what ring theorists coyly refer to as Taylor's lemma, namely

LEMMA. *The map* $x \otimes y \mapsto xy$ *is an isomorphism of* $A^\partial \# k[a]$ *onto* A.

2.2.   The above result obviously requires the present characteristic zero hypothesis. Again given $\sigma \in$ Aut A, call ∂ a σ-derivation of A if $\partial(xy) = (\partial x)y + \sigma(x)\partial y$, $\forall\, x, y \in A$. One can attempt to extend 2.1 to σ-derivations. For example, suppose that there is an invertible central element $q \in A$ such that $\sigma(a) = qa$. Then $\partial a^n = (1 + q + q^2 + \cdots + q^{n-1})a^{n-1}$. We therefore need that every sum $1 + q + q^2 + \cdots + q^m$ be invertible and in particular that q is not a root of unity. Situations in which this can fail occur in graded Lie algebras (with $q = -1$) and quantum groups.

2.3.   Given a locally nilpotent derivation ∂ of a ring A we can form $\exp \partial$ which is an automorphism of A. The situation for a σ-derivation is more complex. First we don't get too far unless σ and ∂ commute. Under this assumption one has

$$\partial^n(xy) = \sum_{m=0}^{n} \binom{n}{m}(\sigma^{n-m}(\partial^m x))(\partial^{n-m}y).$$

Unfortunately the exponents are mixed. Assume further that $\partial x = q^2 x$ with $q \in A$ central and invertible. Then formal summation gives

$$\exp\,\partial(xy) = (\exp\,\partial)x \cdot (\exp\,q\partial)y.$$

In particular if $\partial x = y$, $\partial y = 0$, then

$$(\exp\,\partial)x^n = \prod_{i=1}^{n}(x + q^{2(i-1)}y)$$

so that the q-binomial coefficients make their appearance.
  Indeed set

$$[r]_q = \prod_{s=1}^{t}\left(\frac{q^r - q^{-r}}{q - q^{-1}}\right), \quad \begin{bmatrix} r \\ s \end{bmatrix}_q = \frac{[r]_q}{[r-s]_q[s]_q}.$$

Taking $y = 1$ in the above we obtain

$$(\exp\,\partial)x^n = \prod_{i=1}^{n}(x + q^{2(i-1)}) = \sum_{m=0}^{n} \begin{bmatrix} n \\ m \end{bmatrix}_q x^{m-n}q^{n(m-1)}$$

whilst

$$\partial^r x^s = q^{\frac{1}{2}s(s-1)}\frac{[r]_q}{[r-s]_q}x^{r-s}.$$

Now we can either observe that

$$[\partial^r(x^k(\exp \partial)x^r)]_{x=-1} = (-1)^k q^{2rk}[(\exp \partial)(\partial^r x^r)]_{x=-1} \ ,$$

$$= (-1)^k q^{2rk+\frac{1}{2} r(r-1)}[r]_q$$

or first develop $(\exp \partial)x^r$ in the left hand side which gives

$$\partial^r[x^k(\exp \partial)x^r] = \sum_{s=0}^{r} q^{kr+\frac{1}{2}[r(r-1)-s(s-1)]} \frac{[k+r-s]_q}{[k-s]_q} \begin{bmatrix} r \\ s \end{bmatrix}_q x^{k-s}.$$

Comparison gives the q-analogue

$$\sum_{s=0}^{r}(-1)^s q^{-\frac{1}{2}s(s-1)} \frac{[k+r-s]_q}{[k-s]_q} \begin{bmatrix} r \\ s \end{bmatrix}_q = q^{kr}[r]_q$$

of a well-known binomial identity. (I should like to thank A. Ouakka for checking the computation and in particular obtaining the correct q exponents).

2.4. The above all becomes more interesting when we have several derivations acting simultaneously. Thus suppose $\mathfrak{a} \subset \partial er A$ is a finite dimensional Lie algebra of locally nilpotent derivations of A. (For example take A to be a primitive quotient of the enveloping algebra of a nilpotent Lie algebra $\mathfrak{a}$ which we further view as acting on $\mathfrak{a}$ by derivations). An easy exercise shows that $\mathfrak{a}$ is nilpotent. Then by Engel's theorem applied to $\mathfrak{a}$ we can find a decreasing chain $\mathfrak{a} = \mathfrak{a}_0 \supsetneq \mathfrak{a}_1 \supsetneq \mathfrak{a}_2 \cdots \supsetneq \mathfrak{a}_{n+1} = 0$ of ideals $\mathfrak{a}_i$ of $\mathfrak{a}$ of codimension i. Choose $x_i \in \mathfrak{a}_{i-1} \setminus \mathfrak{a}_i$. Set $A_i = A^{\mathfrak{a}_i}$ which is $\mathfrak{a}$ stable. One easily checks that either $A_{i-1} = A_i$ or there exists $a_i \in A_i$ such that $x_i a_i \in A^{\mathfrak{a}} \setminus \{0\}$. Now suppose that $A^{\mathfrak{a}}$ reduces to scalars (as in the above example). Then we can assume $x_i a_i = 1$ and so by Taylor's lemma $A_{i-1} = A_i \# k[a_i]$. With a little extra work one may show ([29, 3.2]) that it is possible to choose the a_i so that all the commutators $[a_i, a_j]$ are scalars. This gives the

PROPOSITION. *Let $\mathfrak{a}$ be a finite dimensional Lie algebra of locally nilpotent derivations of an associative algebra A. If $A^{\mathfrak{a}}$ reduces to scalars, then A is a polynomial extension of a Weyl algebra.*

2.5. The interest of 2.4 further increases when the set-up is equivariant under some torus action. To be specific, let $\mathfrak{a}$ be a split semisimple Lie algebra with triangular decomposition $\mathfrak{g} = \mathfrak{n} \oplus \mathfrak{h} \oplus \mathfrak{n}^-$. The $\mathfrak{h}$ weight spaces of $\mathfrak{n}$ are one dimensional and the set of weights form a finite subset

$R^+ \subset \mathfrak{h}^*$ with a hyperplane of support through the origin. Now assume A to be a commutative domain on which $\mathfrak{b} := \mathfrak{h} \oplus \mathfrak{n}$ acts by derivations. Assume further that A is a direct sum of its $\mathfrak{h}$ weight spaces A_μ and that its set of weights lie in a finite union of cones $\lambda_i - \mathbb{N}R^+ : \lambda_i \in \mathfrak{h}^*$. This last condition forces the action of $\mathfrak{n}$ on A to be locally nilpotent. Assume finally that $A^{\mathfrak{n}}$ reduces to scalars. In the construction of 2.4 we can obviously take x_i to have weight $\alpha_i \in R^+$ and then if $A_{i-1} \neq A_i$ take a_i to have weight $-\alpha_i$. Set

$$\operatorname{ch} A = \sum_{\mu \in \mathfrak{h}^*} (\dim A_\mu) e^\mu.$$

Since A is a polynomial ring on the x_i a standard computation ([14, 7.5]) shows that

COROLLARY.   *There exists a subset $S \subset R^+$ such that*

$$\operatorname{ch} A = \prod_{\alpha \in S} (1 - e^{-\alpha})^{-1}.$$

2.6.   Retain the hypotheses of 2.5. We can further refine the above example by taking $\mathfrak{m}$ to be an $\mathfrak{h}$ stable subalgebra of $\mathfrak{n}$. Such a subalgebra is determined by its weights which form a subset $T \subset R^+$. Again by 2.4, A is a polynomial extension of $A^{\mathfrak{m}}$ by generators with weights in $T \cap S$. We hence obtain the further

COROLLARY.   *One has*

$$\operatorname{ch} A^{\mathfrak{m}} = \prod_{\alpha \in S \setminus (S \cap T)} (1 - e^{-\alpha})^{-1}.$$

An application of this result will be given in Sects. 8.6 - 8.9.

2.7.   Let M be a $U(\mathfrak{a})$ module. Then $\operatorname{End}_k M$ is a $U(\mathfrak{a})$ bimodule. We have a natural adjoint action of $\mathfrak{a}$ on End M obtained by combining the diagonal action with principal antiautomorphism, namely,

$$(\operatorname{ad} x)\xi := x\xi + \xi\check{x} := x\xi - \xi x.$$

The set $F(M)$ of all endomorphisms on which ad $\mathfrak{a}$ is locally finite is a subring of End M. We may define $F(M, N) \subset \hom_k(M, N)$ analogously. Furthermore by the local finiteness of ad $\mathfrak{a}$ on $U(\mathfrak{a})$, these are bisubmodules. Also the action of $U(\mathfrak{g})$ on M defines a (canonical) homomorphism $U(\mathfrak{a}) \to F(M)$. The study of $F(M)$ plays a fundamental role in enveloping algebras. Notice that we can replace $U(\mathfrak{a})$ by any Hopf algebra $\mathcal{H}$ (with antipode) - thus in particular a quantum group. Then $F(M)$ is a subring; but *not* in general an $\mathcal{H}$ bisubmodule of End M. Nevertheless its study is again of importance in quantum groups. This is because $\mathcal{H}$ admits a large subalgebra $\mathcal{H}_0$ on which adjoint action is locally finite [42]. Moreover this analysis leads to a Harish-Chandra isomorphism for "semisimple" quantum groups [42] which is finer than that report by M. Rosso [61] and T. Tanisaki [65].

2.8.   Assume M simple in 2.7. A basic question is whether the canonical map $U(\mathfrak{a}) \to F(M)$ is surjective. In general $F(M)$ is not even finitely generated over $U(\mathfrak{a})$, however for $\mathfrak{a}$ semisimple this does hold and it is even of finite length as a bimodule ([31, III, 2.4]).

A further interesting devclopment has been to study overrings of $U(\mathfrak{a})/\operatorname{Ann}\,M$ which are locally finite for the diagonal action and finitely generated as left (or right) $U(\mathfrak{a})$ modules. Modifying slightly the terminology of McGovern [53], we call such an overring a Dixmier algebra (over the primitive factor $U(\mathfrak{g})/\operatorname{Ann}\,M$). In this it is usual to assume k algebraically closed which we shall do below.

It is rather easy to classify Dixmier algebras in the nilpotent case. Assume $\mathfrak{g}$ nilpotent and let G denote its algebraic adjoint group. Recall ([10, Chap. 6]) that we have a surjective map Dix : $\mathfrak{g}^* \to$ Prim $U(\mathfrak{g})$ which factors to a bijection of $\mathfrak{g}^*/G$ onto Prim $U(\mathfrak{g})$. Each $\lambda \in (\mathfrak{g}, \mathfrak{g})^\perp$ defines a winding automorphism $\tau_\lambda : x \mapsto x + \lambda(x)$ on $\mathfrak{g}$. Extend τ_λ to $U(\mathfrak{g})$. Then ([10, 11.1]) one has $\tau_\lambda(\operatorname{Dix}(f)) = \operatorname{Dix}(f - \lambda)$. Now fix $P \in$ Prim $U(\mathfrak{g})$ and set $\mathfrak{l}_p = \{\lambda \in (\mathfrak{g}, \mathfrak{g})^\perp \mid \tau_\lambda(P) = P\}$. This is a subgroup of $(\mathfrak{g}, \mathfrak{g})^\perp$ which by the above is rather easy to compute.

It is standard that $[\mathfrak{g}, \mathfrak{g}]^\perp$ also parametrizes the linear representations of $\mathfrak{g}$. We define a k-algebra R to be $\mathfrak{l}_p$ graded if it admits a $\mathfrak{g}$ action by derivations and is a direct sum of its weight spaces $R_\lambda := \{r \in R \mid xr = \lambda(x)r,\ \forall\, x \in \mathfrak{g}\}$ with $\lambda \in \mathfrak{l}_p$. Given such a ring we may form the smash product $U(\mathfrak{g})\#R$ and in which $P \otimes 1$ is two-sided ideal by our assumption on λ. The quotient which may be written $U(\mathfrak{g})/P\#R$, is a Dixmier algebra over $U(\mathfrak{g})/P$ if and only if dim $R < \infty$.

THEOREM.   (*$\mathfrak{g}$ nilpotent*). *Every Dixmier algebra obtains from the above construction.*

Let D be a Dixmier algebra over $U(\mathfrak{g})/P : P \in$ Prim $U(\mathfrak{g})$. Given $\lambda \in [\mathfrak{g}, \mathfrak{g}]^\perp$ we let D_λ denote its λ weight space for the diagonal action of $\mathfrak{g}$ and D^λ the corresponding generalized λ weight space. By local finiteness of diagonal action and the nilpotent Chinese remainder theorem ([14, 1.3.19]) one has a direct sum decomposition

$$D = \oplus D^\lambda.$$

Moreover each D^λ is $U(\mathfrak{g})$ bimodule, so by the finite generation hypothesis, this sum is finite. Suppose $D^\lambda \neq 0$. Then $D_\lambda \neq 0$. Choose $0 \neq d \in D_\lambda$ and let $\ell(d)$ (resp. $r(d)$) denote its left (resp. right) annihilator in $U(\mathfrak{g})$. It is clear that $\ell(d)$ is a two-sided ideal of $U(\mathfrak{g})$ containing P. Yet by 2.4, $U(\mathfrak{g})/P$ is a Weyl algebra over its centre (which reduces to scalars via the Schur and Quillen lemmas [14, 2.6.9] and is hence simple. This proves that $\ell(d) = P$. Similarly $r(d) = P$. Yet $Pd = d(\tau_\lambda(P))$ so we must have $\lambda \in \mathfrak{l}_p$.

Set $A = U(\mathfrak{g})/P$ and $R = \oplus D_\lambda$. Obviously R is a subalgebra of D and is $\mathfrak{l}_p$ graded by the above. Since A is central simple a very slight modification of the usual argument shows that the map $\varphi : a \otimes r \mapsto ar$ is an injection of $A \# R$ into D. Since A is noetherian, the finite generation hypothesis forces R to be finite dimensional.

Finally we show that φ is surjective and this will complete the proof. Here we may fix $\lambda \in \mathfrak{l}_p$ and show that the restriction φ_λ of φ to $A \otimes D_\lambda$ has image D^λ. First assume $\lambda = 0$. We claim that the canonical Weyl algebra generators $\{a_i\}$ of A constructed as in 2.4 act locally nilpotently on D^0. This is not a priori obvious. However by definition, $\mathfrak{g}$ acts locally nilpotently on D^0, thus if we write

$$D^{0,n} = \{a \in D^0 \,\big|\, (\mathrm{ad}\ g)^n a = 0\}, \quad \text{then} \quad D^0 = \cup D^{0,n}.$$

Obviously $(\mathrm{ad}\ \mathfrak{g})D^{0,n} \subset D^{0,n-1}$. Thus viewing a_i as sum of products of the x_j we conclude that $[a_i, D^{0,n}] \subset AD^{0,n-1}$. Yet $\mathrm{ad}\ a_i$ acts locally nilpotently on A and so an easy induction argument proves the required assertion. With this established we can again use Taylor's lemma to prove surjectivity. Notice that this last argument fails at the level of the Lie algebra action because in the notation of 2.4 one must sometimes have $A_{i-1} = A_i$, whilst at the Weyl algebra level this never happens.

A priori the above argument fails for $\lambda \neq 0$. Yet assume $\lambda \neq 0$ and set $\mathfrak{g}_1 = \ker \lambda$ which is an ideal of codimension 1 in $\mathfrak{g}$. Set $A_1 = U(\mathfrak{g}_1)/P \cap U(\mathfrak{g}_1)$. The surjectivity of φ_λ will follow as in case $\lambda = 0$ if we can show that there is a choice of the canonical generators $\{a_i\}$ of A for which the twisted diagonal action $x \mapsto a_i x - x\tau_\lambda(a_i)$ on D^λ is locally nilpotent for all i. Fix $x_0 \in \mathfrak{g} \setminus \mathfrak{g}_1$. Since τ_λ restricts to the identity on A_1 the image of x_0 in A cannot already lie in A_1. This in turn forces $A_1^{x_0} \subsetneq A_1$ for otherwise x_0 would be central in A. By the action of $\mathfrak{g}$ on A_1 it follows from 2.4 that $A_1 = A_1^{x_0} \otimes k[a]$ with $[x_0, a] = 1$ and $A_1^{x_0}$ a Weyl algebra. This shows that we can choose x_0 to be a canonical generator and pick the remaining canonical generators from A_1. Since $\tau_\lambda(x_0) = x_0 + \lambda(x_0)$ the required assertion is clear. The theorem is proved.

2.9.   Had we assumed in 2.8 that the diagonal action of $\mathfrak{g}$ extends to an action of a connected unipotent algebraic subgroup G of $GL(\mathfrak{g}^*)$ with Lie algebra $\mathfrak{g}$, then this would have forced $\mathfrak{g}$ to act locally on D. A further trivialization results if we further either require D to be an integral domain or of the form $F(M)$ with M a simple $U(\mathfrak{g})$ module. In both cases D is just the image of $U(\mathfrak{g})$, the second assertion being an old result of P. Tauvel [67].

With the further restraints of a G action (as above) and integrality, the solvable case becomes tractable. Thus R. Rentschler informs me he was recently able to verify for $\mathfrak{g}$ solvable, algebraic that the Dixmier algebras over $\mathrm{Dix}(f)$ are parametrized by the images of the component group of $\mathrm{Stab}_G f$. Notice here the dependence on the choice of G.

When $\mathfrak{g}$ is semisimple the problem is far more complicated. First of all even when M is a simple highest weight module $F(M)$ one obtains a rich family of Dixmier algebras [43]. Moreover these were crucial to the detailed study of Prim $U(\mathfrak{g})$. D. Vogan [69] suggested that a further filtration hypothesis should be imposed. Under a strong version of this hypothesis C. Moeglin [56] showed that Dixmier algebras occur as subrings of $F(M)$ with M a generalized Whittaker module. A. Zahid [72] and S.P. Smith [62] showed that $F(M)$ can be quite rich even for an ordinary Whittaker module. Finally W. McGovern ([54]), (and references therein) has made an extensive study of Dixmier algebras in the sense of Vogan. Probably like me, ring theorists will abhor such filtration constraints.

§3. Localization

3.1.   Let A be a prime, noetherian ring. By a famous theorem of Goldie, the set S of non-zero divisors of A is an Ore subset. Consequently we can invert the elements of S to form the full quotient ring $S^{-1}A$. By the Artin-Wedderburn theorem $S^{-1}A$ is a matrix ring over a skew field. Its rank is an invariant of A which we call the Goldie rank rk A of A.

A question of immense interest has been the calculation of rk A when A is a primitive quotient of an enveloping algebra of a semisimple Lie algebra $\mathfrak{g}$ over an algebraically closed field of characteristic zero. With respect to a triangular decomposition of $\mathfrak{g}$ each $\lambda \in \mathfrak{h}^*$ defines a unique simple highest weight module $L(\lambda)$ of highest weight $\lambda - \rho$, where ρ is the half sum of the positive roots. A crucial question has been to relate $rk(U(\mathfrak{g})/\mathrm{Ann}\ L(\lambda))$ to

$ch\ L(\lambda)$. Taking $A = U(\mathfrak{g})/\operatorname{Ann}\ L(\lambda)$ and S as above we nearly always have $S^{-1}L(\lambda) = 0$. What we need are much smaller Ore subsets for which $L(\lambda)$ has a chance of being torsion-free. The technical details are too complicated to discuss here; but we analyze below a situation which at least illustrates the principle.

3.2. Let $s \in A$ be locally ad-nilpotent. An easy exercise shows that $S = \{s^k\}_{k \in \mathbb{N}}$ is Ore in A. This is one of the most important examples of small Ore sets.

3.3. Let $\mathfrak{g}$ be a semisimple Lie algebra, $\mathfrak{h}$ a Cartan subalgebra and $R \subset \mathfrak{h}^*$ its set of roots. We can write

$$\mathfrak{g} = \mathfrak{h} \oplus \bigoplus_{\alpha \in R} k x_\alpha$$

where x_α has weight α. Call a $\mathfrak{g}$ module $\mathfrak{h}$-admissable if M is a direct sum of its $\mathfrak{h}$ weights spaces M_μ with dim $M_\mu < \infty$, $\forall \mu \in \mathfrak{h}^*$. Let $\mathcal{F}$ denote the category of all $\mathfrak{h}$ admissable modules. Recently S. Fernando [15] has reduced the classification of simple modules in $\mathcal{F}$ to the case when the action of x_α on M induces an injection $\varphi(x_\alpha) : M_\mu \longrightarrow M_{\mu+\alpha}$, $\forall \mu \in \mathfrak{h}^*$, $\forall \alpha \in R$. Since $\alpha \in R$ implies $-\alpha \in R$ finite dimensionality imposes the bijectivity of each $\varphi(x_\alpha)$. Hence either dim $M_\mu = 0$ or has a value $r(M)$ independent of the choice of μ. We call such modules bijective.

LEMMA. $r(M) = rk(U(\mathfrak{g})/\operatorname{Ann}\ M)$ for any $M \in \operatorname{Ob} \mathcal{F}$ which is bijective.

As we shall see (5.6) the hypothesis of bijectivity implies that $\mathfrak{g}$ has only type A_ℓ or type C_ℓ factors. One then checks that it is possible to find (positive) roots $\beta_1, \beta_2, \cdots, \beta_\ell : \ell = \operatorname{rank} \mathfrak{g}$ such that

(i) $\beta_i + \beta_j$ *is never a root.*

(ii) $\sum_{i=1}^{\ell} \mathbb{Z}\beta_i = \mathbb{Z}R$.

By (i) the root vectors $x_{\beta_i} : i = 1, 2, \cdots, \ell$ span a commutative subalgebra $\mathfrak{m}$ of $\mathfrak{g}$ and hence generate an Ore set S in $U := U(\mathfrak{g})/\operatorname{Ann}\ M$. The hypothesis on M implies the canonical map $M \to S^{-1}M$ is bijective.

Let Q denote the subring of $S^{-1}U$ generated by the $x_{\beta_i}^{\pm 1} : i = 1, 2, \cdots, \ell$ and set $A = QU(\mathfrak{h}) = U(\mathfrak{h})Q$. Let σ denote the collection of automorphisms $\sigma_i : a \mapsto x_{\beta_i} a x_{\beta_i}^{-1}$ of $S^{-1}U$. We claim that the natural maps give isomorphisms

(*) $$(S^{-1}U)^{\mathfrak{h}, \sigma} \otimes A \xrightarrow{\sim} S^{-1}U$$

(**) $$(S^{-1}U)^{\mathfrak{h}, \sigma} \otimes U(\mathfrak{h}) \xrightarrow{\sim} (S^{-1}U)^{\mathfrak{h}}$$

In fact by Taylor's lemma $(S^{-1}U)^{\mathfrak{m}} \otimes U(\mathfrak{h}) \xrightarrow{\sim} S^{-1}U$, whereas by (ii) and the linear independence of the β_i we obtain $(S^{-1}U)^{\mathfrak{h}} \otimes Q \xrightarrow{\sim} S^{-1}U$. Then $(S^{-1}U)^{\mathfrak{m}} = (S^{-1}U)^\sigma \xleftarrow{\sim} ((S^{-1}U)^{\mathfrak{h}} \otimes Q)^\sigma = (S^{-1}U)^{\mathfrak{h},\sigma} \otimes Q$ which gives the required assertions.

By $(**)$, $(S^{-1}U)^{\mathfrak{h},\sigma}$ and $(S^{-1}U)^{\mathfrak{h}}$ have the same restriction to M_μ. Yet the first restricts injectively, whilst one easily checks that M_μ is simple already over $U^{\mathfrak{h}}$. Consequently $(S^{-1}U)^{\mathfrak{h},\sigma}$ identifies with the matrix ring $\mathrm{End}_k M_\mu$. Since A is a localized Weyl algebra and so an integral domain, the required result then obtains from $(*)$.

§4. Filtration

4.1. Let A be a ring with identity 1. By a filtration on A we shall mean an increasing family $0 = A^{-1} \subset A^0 \subset A^1 \subset \cdots$, of additive subgroups such that $A^m A^n \subset A^{m+n}$, $\forall\, m,n \in \mathbb{N}$ and $\bigcup A^n = A$. We always assume $1 \in A^0$.

The first property above implies that

$$\mathrm{gr}\ A = \oplus_{i \in \mathbb{N}} A^i / A^{i-1}$$

admits a ring structure. The general idea is that gr A is an approximation of A which is easier to study; but then that properties of gr A can be lifted to those of A. Two easy results are that gr A being noetherian (resp. integral) implies that A is noetherian (resp. integral).

4.2. A good understanding of filtered and associated graded rings comes from introducing the Rees ring

$$\mathrm{Gr}\ A = \oplus_{i \in \mathbb{N}} A^i.$$

For example, let u denote the identity in A but considered as element of A^1 viewed as a direct summand of Gr A. Then Gr A/u Gr A just identifies with gr A. From this one can get the better result that the noetherianity of gr A and Gr A are equivalent.

Again let M be an A module. A filtration $\mathcal{F}$ of M is an increasing family $0 = \mathcal{F}^{-1}M \subset \mathcal{F}^0 M \subset \cdots$ of additive subgroups such that $A^m(\mathcal{F}^n M) \subset \mathcal{F}^{n+m}M$ and $\bigcup \mathcal{F}^n M = M$. As before we may define

$$\mathrm{gr}_\mathcal{F}\ M = \oplus \mathcal{F}^i M / \mathcal{F}^{i-1}M.$$

Typically one obtains a filtration on M by choosing a generating subspace M^0 and setting $\mathcal{F}^n M = A^n M^0$. Inevitably we need to compare filtrations. Thus we call $\mathcal{F}$, $\mathcal{F}'$ equivalent if there exist $c_1, c_2 \in \mathbb{N}$ such that

$$\mathcal{F}^{n-c_1} M \subset \mathcal{F}'^n M \subset \mathcal{F}^{n+c_2} M,$$

$\forall\ n \in \mathbb{N}$.

Now consider the Rees module defined to be

$$\mathrm{Gr}_{\mathcal{F}} M = \oplus \mathcal{F}^i M$$

which is a graded module over $\mathrm{Gr}\ A$. Observe that $\mathrm{Gr}_{\mathcal{F}} M / u\ \mathrm{Gr}_{\mathcal{F}} M \xrightarrow{\sim} \mathrm{gr}_{\mathcal{F}} M$. We call $\mathcal{F}$ of finite type if $\mathrm{Gr}_{\mathcal{F}} M$ is finitely generated. For example the filtration associated to M^0 has this property if $\dim\ M^0 < \infty$. If $\mathrm{Gr}\ A$ is noetherian, a noetherianity argument shows that any two finite filtrations are equivalent and again that a finite filtration $\mathcal{F}$ on M induces a finite filtration $\mathcal{F}^n N := \mathcal{F}^n M \cap N$ on any submodule.

The importance of equivalence comes from the following

LEMMA.   *Assume $\mathcal{F}, \mathcal{F}'$ equivalent. Then $\mathrm{gr}_{\mathcal{F}} M, \mathrm{gr}_{\mathcal{F}'} M$ admit equivalent normal series.*

Taking intermediate filtrations reduces to case $c_1 = 0$, $c_2 = 1$. The latter means

(i) $u\ \mathrm{Gr}_{\mathcal{F}'} M \subset \mathrm{Gr}_{\mathcal{F}} M \subset \mathrm{Gr}_{\mathcal{F}'} M$,

and further implies

(ii) $u\ \mathrm{Gr}_{\mathcal{F}} M \subset u\ \mathrm{Gr}_{\mathcal{F}'} M \subset \mathrm{Gr}_{\mathcal{F}} M$.

Since $\mathrm{Gr}_{\mathcal{F}'} M / \mathrm{Gr}_{\mathcal{F}} M \xrightarrow{\sim} u\ \mathrm{Gr}_{\mathcal{F}'} M / u\ \mathrm{Gr}_{\mathcal{F}} M$ then (i), (ii) give the required equivalence.

4.3.   From now on we assume that A is defined over a field of characteristic zero and that $\mathrm{gr}\ A$ is commutative. Given $a \in A^n$ we let $\mathrm{gr}_n(a)$ denote its image in A^n / A^{n-1}. Commutativity means that $\mathrm{gr}_{m+n}(a,b) = 0$, $\forall\ a \in A^n$, $b \in A^m$. This gives rise to a Poisson bracket structure on $\mathrm{gr}\ A$ which is defined on homogeneous elements by setting

$$\{\mathrm{gr}_m(a), \mathrm{gr}_n(b)\} = \mathrm{gr}_{m+n-1}[a,b], \ \forall\ a \in A^n, \ b \in A^n.$$

Call an ideal of $\mathrm{gr}\ A$ involutive if it is stable under Poisson bracket. It is obvious that $\mathrm{gr}\ I$ is involutive for any left ideal of A; but this is not so useful. V. Guillemin, D. Quillen and S. Sternberg [23] pointed out that one should ask if its radical $\sqrt{\mathrm{gr}\ I}$ is involutive and proved this in some cases. This used a trace argument necessitating our characteristic zero hypothesis. In a very difficult paper Kashiwara and Kawai [45] established this for a large family of differential operator rings using microlocalization. Fortunately O. Gabber [16] gave a relatively simple and pure algebraic proof establishing the deep

THEOREM.   *Assume* gr A *is commutative noetherian. Then for any left ideal I of A the radical of* gr I *is involutive.*

REMARK.   Under the noetherianity hypothesis $\sqrt{\text{gr } I}$ is a finite intersection of primes P_i. Fix j, choose $a_i \in P_i \backslash (P_i \cap P_j)$ and set $a = \prod_{i \neq j} a_i$. Given $x, y \in P_j$, then xa, $ya \in \sqrt{\text{gr } I}$ and so $\{x, y\} a^2 = \{xa, ya\} - \{x, a\} ya - \{a, y\} xa \in P_j$. Hence $\{x, y\} \in P_j$ by choice of a and the primeness of P_j. Hence P_j is involutive.

4.4.   Assume the hypothesis of theorem 4.3. A very difficult question is to decide when gr I is prime. For I two-sided we must require A/I to be an integral domain; but this is not enough even for A being the enveloping algebra of $\mathfrak{so}(5)$ as noted by M. Duflo (cf. [30, 6.15 remarks]). When I is a prime two-sided ideal or a maximal left ideal, then a further result of O. Gabber [17] asserts that the zero variety of gr I is equidimensional. This is a key step in proving ([9, 33]) that the zero variety of gr P : $P \in \text{Prim } U(\mathfrak{g})$: $\mathfrak{g}$ semisimple, is irreducible. Recently J.E. Björk and E.K. Ekström [7], have shown that when A/I is a simple module one may choose a finite filtration such that gr I has no embedded primes.

4.5.   We can apply the above theory to an enveloping algebra $U(\mathfrak{a})$: dim $\mathfrak{a} < \infty$ with its canonical filtration $\{U^n(\mathfrak{a})\}_{n \in \mathbb{N}}$. We shall further assume that the base field is algebraically closed. One has $\text{gr}(U(\mathfrak{a})) \cong S(\mathfrak{a})$ and this further identifies with the polynomial functions on $\mathfrak{a}^*$. Given a $U(\mathfrak{g})$ module M generated by a finite dimensional subspace M^0, the construction of 4.2 associates a finite filtration $\mathcal{F}$ to M. Then by the lemma $\sqrt{\text{Ann gr}_{\mathcal{F}} M}$ is independent of the choice of M^0. We can then define the associated variety $V(M)$ to be the zero variety in $\mathfrak{a}^*$ of Ann $\text{gr}_{\mathcal{F}} M$. Notice we can do better, namely the multiplicities of each component of $V(M)$ are invariants of M. Let $S(M)$ denote $V(M)$ counted with its multiplicities. From what has been said in 4.2 concerning induced filtrations we obtain the

LEMMA.

   (i) *In top dimension, $S(M)$ is additive on exact sequences.*
   (ii) $S(E \otimes M) = (\dim E) S(M)$ *for any finite dimensional $\mathfrak{a}$ module E.*

4.6.   The above result is a close relative of the Goldie rank additivity principle for $\mathfrak{a} = \mathfrak{g}$ semisimple. Assume M simple. Then $V(M)$ is equidimensional by the remarks in 4.4. Let $\mathfrak{h}$ be a Cartan subalgebra of $\mathfrak{g}$ and M the highest weight module $L(\lambda)$: $\lambda \in \mathfrak{h}^*$. Let us write

$$S(L(\lambda)) = \sum \ell_i(\lambda) [V_i]$$

where the V_i denote the components of $V(L(\lambda))$. To analyze the consequences of 4.5(ii) we may use the same method (see [31, II] or [28]) used to obtain the Goldie rank polynomials obtained from the function $\lambda \mapsto rk\ F(L(\lambda))$. Thus one may show that $\mathfrak{h}^*$ is a disjoint union of Zariski dense subsets Z_λ on which each ℓ_i is the restriction of a (unique) polynomial on $\mathfrak{h}^*$. Moreover each ℓ_i must transform exactly as does the corresponding Goldie rank polynomial under the Weyl group W. Then the simplicity of the Weyl group representations obtained from Goldie rank polynomials implies via Schur's lemma that each ℓ_i is a positive integer scalar multiple c_i of this Goldie rank polynomial. Thus we may write

$$S(L(\mu)) \in \mathbb{Q} \sum c_i(V_i), \quad \forall\, \mu \in Z_\lambda.$$

The above result first obtained in ([32, 5.5]) has a rather nice interpretation. One may assign ([32, Sect. 2]) to any affine conical (or projective) variety V with a torus action, a polynomial p_V which measures the growth rate of the regular (homogeneous) functions on V. In the present case the polynomials p_{V_i} span a module for the Weyl group which is irreducible [27] and the eventual consequence of this is their linear independence. Then an asymptotic analysis of ch $L(\mu)$ shows that $S(L(\mu))$ is proportional to a Goldie rank polynomial and so the above sum is just the decomposition of the polynomial as a linear combination of the p_{V_i}. In general $V(L(\mu))$ is not irreducible and so this sum can have several terms [32, 8.6 - 8.9]. Several authors [9], [12], [38], [68] have found interpretations and results concerning these polynomials.

4.7.   A useful corollary of 4.3 is the following. Let M be a finitely generated $U(\mathfrak{a})$ module, $\mathfrak{b}$ a subspace of $\mathfrak{a}$ which generates $\mathfrak{a}$ as a Lie algebra. Suppose $V(M) \subset \mathfrak{b}^\perp$. Then M is finite dimensional. Indeed since $V(M)$ is involutive, the hypothesis $V(M) \subset \mathfrak{b}^\perp$ implies that $V(M) \subset (\mathfrak{b} + [\mathfrak{b}, \mathfrak{b}] + \cdots)^\perp = \mathfrak{a}^\perp = 0$. Consequently each $x \in \mathfrak{a}$ is algebraic on M. By the hypothesis of finite generation this implies the required result. Notice that this result is false for an infinite dimensional Kac-Moody Lie algebra $\mathfrak{g}$, since it false for the adjoint representation. In this case the noetherianity hypothesis needed in 4.3 is not satisfied if we take A to be the enveloping algebra of $\mathfrak{g}$.

4.8.   There is a more subtle application of 4.3 which can be applied to study the category $\mathcal{F}$ of modules introduced in 3.3.

THEOREM.   *Each $M \in \mathrm{Ob}\,\mathcal{F}$ which is finitely generated, has finite length.*

Recall the notation 3.3. Without loss of generality we can assume M to be generated by a finite sum of weight spaces V. Let $Z(\mathfrak{g})$ denote the centre of $U(\mathfrak{g})$. Then $Z(\mathfrak{g})V \subset V$ and so $\mathrm{Ann}_{Z(\mathfrak{g})} M = \mathrm{Ann}_{Z(\mathfrak{g})} V$ has finite codimension. We can therefore assume that $Z(\mathfrak{g})$ sets by scalars without loss of generality.

Now take any root $\alpha \in R$ and consider $x_\alpha x_{-\alpha}$. Since $(x_\alpha x_{-\alpha})V \subset V$ and V is finite dimensional, there exists a polynomial p such that $p(x_\alpha x_{-\alpha})V = 0$. Taking the filtration of M defined by $\mathcal{F}^n M = U^n(\mathfrak{g})V$ it easily follows that $x_\alpha x_{-\alpha} \in \sqrt{\mathrm{Ann}\ \mathrm{gr}_{\mathcal{F}} M}$. Let C be an irreducible component of $V(M)$ and $I(C)$ its ideal of definition in $S(\mathfrak{g})$. Set $R_C = \{\alpha \in R \mid x_\alpha \in I(C)\}$. By the above for each $\alpha \in R$ either $\alpha \in R_C$ or $-\alpha \in R_C$. Now take $\alpha, \beta \in R_C$ with $\alpha + \beta \in R$. By 4.3 and the remark, $I(C)$ is involutive and so $x_{\alpha+\beta} = \{x_\alpha, x_\beta\} \in I(C)$ hence $\alpha + \beta \in R_C$. It is an easy exercise to show that these two properties of R_C imply that there exists a positive system R^+ such that $R_C \supset R^+$. One easily checks (as above) that $\mathfrak{h} \subset I(C)$ and so if $\mathfrak{b}$ is Borel subalgebra containing $\mathfrak{h}$ corresponding to R^+ we conclude that $\mathfrak{b} \subset I(C)$.

Now identify $\mathfrak{g}^*$ with $\mathfrak{g}$ through the Killing form. Let G denote the adjoint group of $\mathfrak{g}$ and B the subgroup corresponding to $\mathfrak{b}$. By the above C is an irreducible B stable involutive subvariety of $\mathfrak{n} = \mathfrak{b}^\perp$. As noted in say ([32, Sect. 7]) one has a complete classification of such varieties. We call such varieties, orbital varieties. In particular one has dim $GC = 2$ dim C.

Now let $V(\mathrm{Ann}\ M)$ denote the associated variety of $U(\mathfrak{g})/\mathrm{Ann}\ M$ considered as a left $U(\mathfrak{g})$ module. An elementary calculation shows that $V(\mathrm{Ann}\ M) \supset GV(M)$ and again it is a general fact (5.5) that dim $V(\mathrm{Ann}\ M) \leq 2$ dim $V(M)$, for $\mathfrak{g}$ algebraic. Now dim $V(\mathrm{Ann}\ M) \geq$ dim $GV(M) = \sup_i$ dim $GC_i = 2 \sup_i$ dim $C_i = 2$ dim $V(M)$ so in fact we have the equality

$$\mathrm{dim}\ V(\mathrm{Ann}\ M) = 2\ \mathrm{dim}\ V(M),$$

for any $M \in \mathrm{Ob}\,\mathcal{F}$ which is finitely generated. This and because we can assume $Z(\mathfrak{g})$ acts by scalars, implies via ([37, 2.2.9]) that M has finite length.

4.9.   A significant open problem is to show that each orbital variety C is the associated variety of a simple highest weight module. Discussion of this question may be found in [32, Sect. 8] and [21], [22]. A further refinement is discussed in [36, Sect. 8]. Recently E. Benlulu [4] has shown that conjective ([36, 8.4]) fails even in $\mathfrak{sl}(6)$; yet it is probably that a natural modification will still hold. A variant of this question for Harish-Chandra modules has been studied by Vogan [71]. It gives in particular a new proof of the main result of [33].

§5. Growth rates

5.1.   Let A be a k-algebra generated by a finite dimensional subspace U which we assume contains the identity. Let M be an A module generated by a finite dimensional subspace V. An easy exercise shows that

$$\overline{\lim}_{n\to\infty} \frac{\log \dim_k U^n V}{\log n}$$

is independent of the choice of generating subspace U, V. It is called the Gelfand-Kirillov dimension $d_A(M)$ of M over A. Set $d(A) = d_A(A)$, that is when A is viewed as a module over itself. It is not difficult to extend the definition of $d_A(M)$ in the absence of finite generation.

5.2.   The use of Gelfand-Kirillov dimension (or similar growth rate estimates) has made some difficult looking problems quite easy. We give a few striking examples below. The reader may also consult [47] and [52, Chap. 8].

5.3.   One of the earliest uses of Gelfand-Kirillov dimension which can be found in ([11, 6.1]) was developed from an argument which even preceded the invention of this concept! Observe that $d(A) \geq d(B)$ if B is a subring of A. One easily checks that $d(B)$ is infinite if B contains a free subalgebra on two or more generators. Again $d(U(\mathfrak{g})) = \dim \mathfrak{g} < \infty$ for a finite dimensional Lie algebra $\mathfrak{g}$. Thus any two elements $a, b \in U(\mathfrak{g})$ must satisfy a relation of the form $p(a, b) = 0$ where p is a (non-commuting) polynomial. Now assume a, b non-zero and p of minimal degree > 0. We can write $p(a, b) = r(a, b)a + s(a, b)b$ where r, s are similar polynomials but of strictly lower degree than p. Since $U(\mathfrak{g})$ is an integral domain, both must be non-vanishing. This argument proves that *any* subalgebra of $U(\mathfrak{g})$ admits a classical skew-field of quotients.

5.4.   Obviously $d_A(M)$ is an invariant so it can be used to distinguish isomorphism classes. For example if $A_n = k(x_1, x_2, \cdots, x_n, \frac{\partial}{\partial x_1}, \cdots, \frac{\partial}{\partial x_n})$ denotes the Weyl algebra of index n then $d(A_n) = 2n$ and so Weyl algebras of differing index cannot be isomorphic.

5.5.   Let M be a non-zero module over the Weyl algebra A_n. It is a classical fact that $\dim M < \infty$ only if $n = 0$. I.N. Bernstein generalized this to show that $d_{A_n}(M) \geq n$. His original argument was rather messy and long. In fact an easy argument shows that $2d_{A_n}(M) \geq d_{A_n}(A_n/\operatorname{Ann} M)$. Not only does this hold in all characteristic; but A_n can be replaced for example by $U(\mathfrak{g})$ for any finite dimensional nilpotent Lie algebra $\mathfrak{g}$. O. Gabber further modified the analysis to show that one can take $\mathfrak{g}$ to be algebraic ([47, 9.11]). We remark here that $d_{U(\mathfrak{g})}(M) = \dim V(M)$ and so this result can be cast in the form $\dim V(\text{Ann } M) \leq 2 \dim V(M)$ which we used in 4.8. Actually since $V(M)$ is an involutive subvariety of $V(\text{Ann } M)$ we obtain this result whenever $V(\text{Ann } M)$ is a finite union of G orbits (notation 4.8). In particular if $V(\text{Ann } M)$ is equidimensional, then this inequality holds for every component C of $V(M)$. Specifically if we return to a finitely generated module M over A_n with char $k = 0$ (which implies that A_n is simple) then every component C of $V(M)$ satisfies $\dim C \geq n$.

5.6.   Assume $\mathfrak{g}$ split semisimple. If a finitely generated $U(\mathfrak{g})$ module M admits a formal character ch M (cf. 2.5) then $d_{U(\mathfrak{g})}(M)$ can be computed from the asymptotics of $\dim M_\mu : \mu \in \mathfrak{h}^*$. Indeed let (,) denote the Cartan inner product on $\mathfrak{h}^*$. By the finite generation hypothesis on M it follows that for each $n \in \mathbb{N}$,

$$\Omega_n := \{\mu \in \mathfrak{h}^* \,|\, (\mu, \mu) \leq n, \quad M_\mu \neq 0\}$$

is a finite set. Moreover one easily checks that

$$d_{U(\mathfrak{g})}(M) = \overline{\lim}_{n \to \infty} \log\Big(\sum_{\mu \in \Omega_n} \dim M_\mu\Big)/\log n.$$

Now assume $M \in \text{Ob}\,\mathcal{F}$ is simple and bijective (3.3). A nice observation of S. Fernando [15] is that such modules only exist when $\mathfrak{g}$ has just A_n or C_n factors. This obtains as follows. For such a module it is immediate from the previous relation that

$$d_{U(\mathfrak{g})}(M) = rk\ \mathfrak{g}.$$

Then the equality noted in 4.8 gives

$$\dim V(\text{Ann } M) = 2\ rk\ \mathfrak{g}.$$

Yet $Z(\mathfrak{g})$ acts by scalars on M and so $V(\text{Ann } M)$ is contained in the cone of nilpotent elements of $\mathfrak{g}^*$ (identified with $\mathfrak{g}$) and a finite union of nilpotent orbits. (One now knows that every $V(P) : P \in \text{Prim } U(\mathfrak{g})$ is the closure of just one nilpotent orbit; but this is much harder [33, 3.10]). Thus $\mathfrak{g}^*$ must admit at least one nilpotent orbit of dimension $2\ rk\ \mathfrak{g}$. This can only happen if $\mathfrak{g}$ has only type A_n or C_n factors.

5.7.   One may make more careful growth rate estimates in special cases. Thus if A is an enveloping algebra then (notation 5.1) one may show that

$$p(n) := \dim_k U^n V$$

is a polynomial for n sufficiently large whose degree d obviously coincides with $d_A(M)$. Now if we write

$$p(n) = \frac{e\,n^d}{d!} + O(n^{d-1})$$

then e is a positive integer (because $p(n)$ takes integer values). Moreover (for fixed U) e is independent of the choice of generating subspace V. We may view it as an invariant $e(M)$ of M, called the Bernstein multiplicity of M. A consequence of all this is the following lemma, which fails ([52, 8.3.4]) for arbitrary rings. Set $d = d_{U(\mathfrak{g})}$.

LEMMA.   Let M be a finitely generated $U(\mathfrak{g})$ module.

  (i) $d(M) = \max\{d(N), d(M/N)\}$ for any submodule N of M. Moreover if all three integers coincide, then $e(M) = e(N) + e(M/N)$.

  (ii) Let $M = M_1 \supseteq M_2 \supseteq \cdots$, be an infinite decreasing chain of submodules of M. One has $d(M_i/M_{i-1}) < d(M)$ for all i sufficiently large.

An immediate consequence of this result is that for a finitely generated module M over the Weyl algebra A_n (over a field of characteristic zero) one has

$(*)$ $d(M) = n \Longrightarrow M$ has finite length.

More generally $K \dim\ M \leq d(M) - n$ where $K \dim$ denotes Krull dimension in the sense of Gabriel-Rentschler ([52, Chap. 6]). A more sophisticated version of the above argument using differential operators on the flag variety shows that for $\mathfrak{g}$ semisimple k algebraically closed and $P \in \mathrm{Prim}\ U(\mathfrak{g})$ one has

$$K \dim\ U(\mathfrak{g})/P \leq \frac{1}{2}(\dim\ \mathfrak{g} - rk\ \mathfrak{g}).$$

The opposite equality is known to hold [48] for P minimal. We remark that the converse of $(*)$ fails. An A_n module satisfying $(*)$ is called holonomic.

5.8.   Actually Bernstein made a much cleverer use of $e(M)$ to show that for any polynomial p in $x_1, x_2, \cdots, x_n$ the localized module $k[x_1, x_2, \cdots, x_n, p^{-1}]$ over A_n which is not obviously finitely generated has in fact finite length (see 7.9). This proved the existence of a differential opertor $\partial \in A_n$ such that $\partial p^n = b(n)p^{n-1}$ for some polynomial b, called the Bernstein polynomial. When $k = \mathbb{C}$, this result has some important applications to analytic continuation and existence of fundamental solutions of the differential equation obtained by interpreting p as a constant coefficient differential operator.

We may adapt this finer aspect of Bernstein's analysis to study the following question. Assume $\mathfrak{g}$ semisimple over an algebraically closed field of characteristic zero. With respect to a triangular decomposition (2.5) let L be a simple highest weight module and recalling 2.7 set $A(L) = \{\xi \in \operatorname{End}_k M \mid \dim(U(\operatorname{ad} \mathfrak{n}^-)\xi) < \infty\}$. (Of course we can more generally define $A(M, N)$). Obviously $A(L)$ contains $F(L)$; but does one have equality? In fact it holds rather often and this is a significant generalization of a remarkable surjectivity result of T. Levasseur, S. Smith and T. Stafford [49]. The proof uses in particular this fine analysis of Bernstein to show that $A(L)$ is a Goldie ring. To see why this is important, let us use $\mathfrak{u}$ to denote ad $\mathfrak{n}^-$. It is immediate from the definition of $A(L)$ that

$$G(L) := \bigcap_{i=0}^{\infty} \mathfrak{u}^i A(L)$$

is a two-sided ideal of $A(L)$. Moreover a little knowledge of the Bernstein-Gelfand-Gelfand $\mathcal{O}$ category shows that $A(L) = F(L) + G(L)$. Necessarily $A(L)$ is a primitive ring and so one may use Goldie's theorem with an elementary Gelfand-Kirillov dimension estimate to show that either $A(L) = F(L)$ or $F(L) \cap G(L) \neq 0$. The latter condition has the following remarkable consequence [40]. There exists a finite dimensional module E such that L is a submodule of $E \otimes L'$ with L' a simple highest weight module induced from a proper parabolic subalgebra of $\mathfrak{g}$. This is a rather rare situation as it puts a severe constraint on $V(L)$ (which equals $V(L')$ via the analysis of 4.5). In particular it never holds if $V(L)$ is properly contained in the nilradical of a maximal parabolic subalgebra of $\mathfrak{g}$.

All this leads to the following conjecture. Let $\operatorname{Prim}_c U(\mathfrak{g})$ denote the set of completely prime primitive ideals of $U(\mathfrak{g})$, that is those for which the quotient algebra is integral. Call $M \in \operatorname{Ob}\mathcal{O}$ quasi-simple if it admits a unique simple submodule M_0 such that $d(M/M_0) < d(M)$.

CONJECTURE.   Take $P \in \operatorname{Prim}_c U(\mathfrak{g})$. Then there exists a quasi-simple $M \in \operatorname{Ob}\mathcal{O}$ such that $P = \operatorname{Ann} M$ and $A(M)$ is a simple ring.

(Probably we should not force M to be simple in the above).

Given the truth of such a result we would either have $F(M) = A(M)$ or $G(M) = A(M)$. In the first case $F(M)$ is itself simple. Under the hypothesis on M one has (cf. 5.9) $U(\mathfrak{g})$ bimodule maps $F(M) \hookrightarrow F(M_0, M) \xrightarrow{\sim} F(M_0)$. By ([41, 6.6]) Soc $F(M_0)$ is an ideal of $F(M_0)$ from which we conclude that $F(M)$ is semisimple as a $U(\mathfrak{g})$ bimodule. This further implies that $U(\mathfrak{g})/\operatorname{Ann} M$ is a simple ring and hence that $\operatorname{Ann} M$ is a maximal ideal. One may then go on to show ([40, 3.5]) that $V(\operatorname{Ann} M)$ is the closure of a rigid orbit (here rigid means non-induced but in a sense which is a bit stronger than the usual one).

In the second case $G(M) = A(M)$. This means that the trivial ad $\mathfrak{g}$ module $\operatorname{End}_{\mathfrak{g}} M$ occurring in $F(M)$ is non-trivially extended in $A(M)$. It implies that M itself is non-trivially induced from a proper parabolic subalgebra. Consequently we can recover all of $\operatorname{Prim}_c U(\mathfrak{g})$ by constructing those ideals associated with rigid orbits and inducing. Actually for the reduction argument to go through we need all to assume that $A(M)$ can be chosen to be integral in the conjecture; but this is probably not a serious requirement. Thus the natural choice for M is $M(\lambda)/PM(\lambda)$ where $M(\lambda)$ is the projective Verma module satisfying $\operatorname{Ann}_{Z(\mathfrak{g})} M(\lambda) = Z(\mathfrak{g}) \cap P$. We note below that this forces $rk\, A(M) = rk(U(\mathfrak{g})/\operatorname{Ann} M) = 1$.

5.9.   Retain the hypothesis of 5.8. Let M_1, M_2 be simple highest weight modules. One has $F(M_1, M_2) = 0$ unless $d(M_1) = d(M_2)$. This is an easy exercise in G.K. dimension which we omit. It was used implicitly in 5.8 to get the embedding $F(M) \hookrightarrow F(M_0)$. The corresponding assertion for $A(M_1, M_2)$ and the relation between $A(M_0)$ and $A(M)$ is more subtle. To study this let S denote the (Ore) set of regular elements of $U(\mathfrak{g})/\operatorname{Ann} M_0$. Set $A = A(M)$, $A_0 = A(M_0)$. Recall ([28, 8.14]) that $\operatorname{Ann} M = \operatorname{Ann} M_0$ and that $S^{-1}A_0$ identifies with Fract A_0 via ([36, 7.11]).

PROPOSITION.   *Assume $M \in \operatorname{Ob}\mathcal{O}$ is quasi-simple with $M_0 = \operatorname{Soc} M$. Then $aM_0 = 0 : a \in A(M)$ implies $a = 0$. Furthermore the restriction map $A(M) \longrightarrow A(M_0, M)$ induces an embedding of A into $S^{-1}A_0$. In particular Fract A = Fract A_0 = Fract $F(M_0)$.*

For the first part it is enough to show that $B := A(M/M_0, M) = 0$. If $B \neq 0$, then $B(M/M_0) \supset M_0$. It follows that we can find a finite dimensional subspace H of B such that $H(M/M_0)$ contains the canonical generator e_0 of M_0. Moreover we can assume H to be $\mathfrak{u}$ stable by the definition of $A(\cdot, \cdot)$. Then $H(M/M_0) \supset U(\mathfrak{n}^-)e = M_0$. Yet as an $\mathfrak{n}^-$ module $H(M/M_0)$ identifies with an image of $H \otimes M/M_0$ and so

$$d_{U(\mathfrak{n}^-)}(M_0) \leq d_{U(\mathfrak{n}^-)}(H \otimes M/M_0) \leq d_{U(\mathfrak{n}^-)}(M/M_0)$$

Yet $d_{U(\mathfrak{g})}, d_{U(\mathfrak{n}^-)}$ coincide on $\mathcal{O}$ and so this contradicts the quasi-simplicity of M.

Let B be the left S-torsion $U(\mathfrak{g})$ submodule of A. If $B \neq 0$, then $BM \supset M_0$. It follows that we can find a finite dimensional subspace H

of B such that HM contains the canonical generator e_0 of M_0. Moreover we can assume H to be $\mathfrak{u}$ stable without loss of generality. Then $M_0 \subset U(\mathfrak{n}^-)HM = HU(\mathfrak{n}^-)M = HM$. Yet since H is finite dimensional we can choose $s \in S$ such that $sH = 0$. Then $sM_0 = 0$, contradicting that M_0 is a faithful $U(\mathfrak{g})/\operatorname{Ann} M$ module. Hence $B = 0$.

By the first part the restriction map $A(M) \xrightarrow{i} A(M_0, M)$ is bijective. Set $J = \operatorname{Ann}_{U(\mathfrak{g})}(M/M_0)$. Then $J(i(A)) \subset A_0$. Yet $d(M/M_0) < d(M)$ implies $J \neq 0$ and so $J \cap S \neq \emptyset$ by Goldie's theorem. By the second part we obtain the required embedding $A = i(A) \hookrightarrow S^{-1}A_0$. Finally $S^{-1}A(M_0, M/M_0) = 0$ since already $JA(M, M/M_0) = 0$. Hence $S^{-1}A \xrightarrow{\sim} S^{-1}A_0$ as required.

REMARKS.   It is false that $A(M, M/M_0) = 0$. In fact one can easily arrange for M to be induced from a proper parabolic subalgebra whilst M_0 is not. Then the trivial ad $\mathfrak{g}$ module is non-trivially extended in $A(M)$ but not in $A(M_0)$. For example take $\mathfrak{g}$ simple of type A_2 and $P = \operatorname{Ann} L(s_\alpha \rho) : \alpha$ simple. Then $M := M(\rho)/PM(\rho)$ is induced from the parabolic subalgebra $\mathfrak{p}$ defined by the second root, whilst $M_0 = L(s_\alpha \rho)$ is not an induced module. Let us further show that in this case $A(M_0)$ is not noetherian. A standard computation ([30, Sect. 5]) shows that M identifies with the standard module $k[x_{-\alpha}, x_{-(\alpha+\beta)}]$ for the Weyl algebra $A_2 := k[y_\alpha, y_{\alpha+\beta}, x_{-\alpha}, x_{-(\alpha+\beta)}] : y_\alpha = \partial/\partial x_{-\alpha}, \ y_{\alpha+\beta} = \partial/\partial x_{-(\alpha+\beta)}$. Here the subscripts designate ad $\mathfrak{h}$ eigenvalues with respect to the embedding of $U(\mathfrak{g})/\operatorname{Ann} M$ in A_2. Furthermore M_0 is just the augmentation ideal of M viewed as a polynomial ring.

Let $\mathfrak{m}^-$ be the $\mathfrak{h}$ stable complement to $\mathfrak{p}$ in $\mathfrak{g}$. Since the Levi factor of $\mathfrak{p}$ acts locally finitely in M, M_0 it follows that $A(\cdot, \cdot)$ is just the ad $\mathfrak{m}^-$ locally finite part of $\hom_k(\cdot, \cdot)$. Again $\mathfrak{m}^- = kx_{-\alpha} \oplus kx_{-(\alpha+\beta)}$ and so $A(M)$ coincides with A_2. Using say (2.2) one carefully checks that one has an isomorphism $A(M) \xrightarrow{\sim} A(M_0, M)$. It is then clear that $A(M_0)$ identifies with the $\mathfrak{h}$ stable complement to the augmentation ideal I of $k[y_\alpha, y_{\alpha+\beta}]$. The afficiendos will recognize that this is not a noetherian ring. Notice that I viewed as an ad $\mathfrak{g}$ module extends the trivial representation.

An obvious question that arises is why does this not all occur already in type A_1. The same identifications give $M = k[x_{-\alpha}]$, $A(M) = k[y_\alpha, x_{-\alpha}]$ and M_0 the augmentation ideal of $k[x_{-\alpha}]$. In this case the embedding $A(M) \hookrightarrow A(M_0, M)$ is not an isomorphism since multiplication by $x_{-\alpha}^{-1}$ does not occur in the image showing that one cannot be too careful. Then $A(M_0)$ is generated by $x_{-\alpha}$ and $y_\alpha - x_{-\alpha}^{-1}$ which is again a Weyl algebra.

5.10.   It is perhaps worthwhile to emphasize the following result obtained from the above analysis. Let $L_1, L_2 \in \operatorname{Ob}\mathcal{O}$ be simple. If $d(L_1) < d(L_2)$, then $A(L_1, L_2) = 0$, whilst $A(L_2, L_1)$ need not be zero. In particular if L_2 is a Verma module, which we recall is a free rank one $U(\mathfrak{n}^-)$ module, we obtain a vector space isomorphism $A(L_2, L_1)^{\mathfrak{n}^-} = \hom_{\mathfrak{n}^-}(L_2, L_1) \xrightarrow{\sim} L_1$

5.11. The conjecture in 5.8 has an important special case which is worth singling out. Suppose that the associated variety of $P \in \mathrm{Prim}_c\, U(\mathfrak{g})$ is the closure of a rigid orbit $\mathcal{O}$. Then one cannot find M as in the conjecture such that $G(M) = A(M)$ for otherwise M would be non-trivially induced and hence so would $GV(\mathrm{Ann}\ M) = \bar{\mathcal{O}}$. Thus if $A(M)$ is simple we must have $A(M) = F(M)$ and hence (as discussed in 5.8) that Ann $M = P$ is a maximal ideal. This is a very strong conclusion; but nevertheless there are so far no known counterexamples.

§6. Reduction to positive characteristic

6.1. Several famous problems in enveloping algebras have been solved by reduction to positive characteristic. Let us quickly mention three and then turn to a quite unknown problem of my own solved by L. Makar-Limanov using such reduction. This is devastingly simple and perhaps represents the only reasonable proof.

6.2. The Kazhdan-Lusztig conjectures [3] were reduced by a number of equivalences of categories until the coup de grâce could be applied using the Frobenuis map via the Lefshetz theorem. This is hardly within our present scope though we shall briefly discuss the first of these equivalences. Alternative proofs have being given using Hodge theory but this is hardly any simpler.

6.3. H.H. Andersen [2] established the correctness of the Demazure character formula using the very fine Steinberg tensor product formula which has no analogue in characteristic zero.

6.4. The correctness of the Demazure character formula was also established by using Frobenuis splitting. More recently O. Mathieu [55] has used Frobenuis splitting to get refined results of certain filtration conjectures which are very hard [57] to obtain without this method.

6.5. (char $k = 0$). Now consider the following elementary question. Let $\mathfrak{g}$ denote the five dimensional solvable Lie algebra $\mathfrak{g}$ with basis x_1, y_1, x_2, y_2, z with the only non-zero Lie brackets being

$$(1) \qquad\qquad [x_1, y_1] = y_1,\ [x_2, y_2] = y_2,\ [x_1, x_2] = z.$$

One easily checks that $Z(\mathfrak{g}) = k[z]$. Although $\mathfrak{g}$ is not an algebraic Lie algebra one can still ask if Fract $U(\mathfrak{g})$ is isomorphic to the Weyl skew field of index 2 over $k(z)$ which we denote by $D_2(z)$. This is certainly almost true since in Fract $U(\mathfrak{g})$ we can rewrite the above relations in the form

$$(2) \qquad [q_1, p_1] = 1, \quad [q_2, p_2] = 1, \quad [q_1, q_2] = p_1^{-1} p_2^{-1} z$$

where $q_i = y_i^{-1} x_i$, $p_i = y_i : i = 1, 2$. The question is how to get rid of the term $p_1^{-1} p_2^{-1} z$? The obvious method is to replace q_i by $q_i + g_i(p_1, p_2) z$: $i = 1, 2$, for some rational functions g_1, g_2. This leads to the differential equation

$$(3) \qquad \frac{\partial g_2}{\partial p_1} - \frac{\partial g_1}{\partial p_2} = p_1^{-1} p_2^{-1}$$

which unfortunately has no rational solutions.

To in fact prove that one has no such isomorphism, it is enough to prove this fails for infinitely many specializations where z is replaced by a (positive) integer. The calculation is the same in all cases and we just take $z = 1$. The Weyl field D_2 of course corresponds to the choice $z = 0$. Now let ℓ be any prime (in general that does not divide the value assigned to z). Let E (resp. D) denote the skew field over $\overline{\mathbb{F}}_\ell$ with generators defined by (1) with $z = 1$ (resp. $z = 0$) Let $Z(E)$ (resp. $Z(D)$) denote the centre of E (resp. D). Set $a_i = x_i(x_i + 1) \cdots (x_i + \ell - 1) : i = 1, 2$. Following a suggestion of L. Makar-Limanov we first prove the

LEMMA.

    (i) $Z(D) = \overline{\mathbb{F}}_\ell(a_1, a_2, y_1^\ell, y_2^\ell)$. In particular $\dim_{Z(D)} D = \ell^4$.
    (ii) $Z(E) = \overline{\mathbb{F}}_\ell(a_1^\ell, a_2^\ell, y_1^\ell, y_2^\ell)$. In particular $\dim_{Z(E)} E = \ell^6$.

(i) is well known so we omit the proof which in any case is similar to (ii). We note that $(y_i^{-1} x_i)^\ell = a_i y_i^{-\ell} : i = 1, 2$.

(ii) Obviously $[x_1, y_1^\ell] = \ell y_1^\ell = 0$. Thus $y_1^\ell \in Z(E)$. Similarly $y_2^\ell \in Z(E)$. From $x_1 y_1 = y_1 x_1 + y_1 = y_1(x_1 + 1)$ one checks that $[a_1, y_1] = 0$. Since $[(x_1 + t)^\ell, x_2] = \ell(x_1 + t)^{\ell-1} = 0, \ \forall \ t \in \overline{\mathbb{F}}_\ell$ we conclude that $a_1^\ell \in Z(E)$. Similarly $a_2^\ell \in Z(E)$. To prove that the given elements generate the centre we first show that $A := \overline{\mathbb{F}}_\ell[x_1, x_2, y_1, y_2]$ admits a simple ℓ^3 dimensional module M.

To construct M we start from a suitable simple $\mathfrak{g}$ module. Define $f \in \mathfrak{g}^*$ by $f(x_i) = 0$, $f(y_i) = 1$, $f(z) = 1 : i = 1, 2$. One checks that $\mathfrak{p} := ky_1 + ky_2 + kz$ is a Vergne polarization for f and so the induced module $N := U(\mathfrak{g}) \otimes_{U(\mathfrak{p})} k_f$ is simple ([10, 9.5]). This may also be checked directly (as below). Identify N (via the symmetrization map [13, Sect. 5] with the polynomial ring $k[x_1, x_2]$. Then we have a homomorphism

$\varphi : U(\mathfrak{g}) \longrightarrow \text{End}_k N$ defined by $\varphi(x_1)x_1^i x_2^j = x_1^{i+1} x_2^j$, $\varphi(x_2)x_1^i x_2^j = -i x_1^{i-1} x_2^j + x_1^i x_2^{j+1}$, $\varphi(y_1)x_1^i x_2^j = (x_1 - 1)^i x_2^j$, $\varphi(y_2)x_1^i x_2^j = x_1^i (x_2 - 1)^j$, $\varphi(z) = 1 d_N$.

Since the action of $\mathfrak{g}$ is defined over $\mathbb{Z}$ we may replace the base field k by $\bar{F}_\ell$. The resulting A module N_1 is no longer simple. It admits a submodule N_2 of codimension ℓ^3 generated by $a_1^\ell a_2$. Set $M = N_1/N_2$. We claim that M is the required simple module. It has first the correct dimension. Secondly, the action of x_1, x_2 imply that it admits 1 as a cyclic vector. The action of y_2 implies that any non-zero submodule M_0 contains a non-zero polynomial in x_1. By the action of x_1 we can assume that the polynomial is $x_1^{\ell^2 - 1}$. Now assume that $x_1^j \in M_0$ for $\ell(\ell-1) < j \leq \ell^2 - 1$. Combining the actions of x_2, y_2 we deduce that $-j x_1^{j-1} \in M_0$ so we can assume $x_1^{\ell(\ell-1)} \in M_0$. Finally assume that $p_j(x_1^\ell) \in M_0$ for some monic polynomial p_j of degree $j : 0 < j \leq \ell - 1$. Since $\varphi(y_1)p_j(x_1^\ell) = p_j((x_1 - 1)^\ell) = p_j(x_1^\ell + (-1)^\ell) = (j - 1) + O(x_1^{\ell(j-2)})$, we obtain a monic polynomial p_{j-1} of degree $j - 1$ such that $p_{j-1}(x_1^\ell) \in M_0$. Hence $1 \in M_0$ and so $M_0 = M$ proving the simplicity of M. (This seems to be a little harder than the proof of simplicity of N in characteristic zero).

Now consider $Z_0 := \bar{F}_\ell[a_1^\ell, a_2^\ell, y_1^\ell, y_2^\ell]$ Obviously Z_0 is contained in the centre of $Z(A)$ and A is a free rank ℓ^6 module over Z_0 with generators $x_1^{i_1} x_2^{i_2} y_1^{j_1} y_2^{j_2} : 0 \leq i_1, i_2 < \ell^2 - 1$, $0 \leq j_1, j_2 < \ell$. By Schur's lemma $Z(A)$ maps to scalars under φ and by the Jacobson density theorem the map $\varphi : A \to \text{End } M$ is surjective. This proves that A has rank $\geq \ell^6$ as a $Z(A)$ module and so $Z(A) = Z_0$.

To complete the proof we show that $Z(E) = \text{Fract } Z(A) =: K$. Given $z \in Z(E)$, set $I = \{a \in A \mid az \in A\}$. Obviously I is a non-zero two-sided ideal of A. It is enough to show that $I \cap Z(A) \neq 0$, equivalently that $\hat{I} := I \otimes_{Z(A)} K = \hat{A} := A \otimes_{Z(A)} K$.

Consider $\hat{I}$ as a module for the adjoint action of $\mathfrak{g}$. By a version of Lie's theorem in positive characteristic (6.9) there exists $\lambda \in \mathfrak{g}^*$ such that the generalized weight subspace $\hat{I}^\lambda$ of $\hat{I}$ is non-zero. Choose $a \in \hat{I}^\lambda$. Since y_1, y_2 commute, are ad $\mathfrak{g}$ eigenvectors and act locally ad-nilpotently on $\hat{A}$ we can assume that a commutes with y_1, y_2. Now it is obvious that any $\hat{A}$ weight space $\hat{A}_\mu : \mu \in \mathfrak{g}^*$ is at most 1 dimensional over K (and in this case a monomial in y_1, y_2). Furthermore $\hat{A}^\mu = \hat{A}_\mu \hat{A}^0$. Thus writing $b = az$ it is clear that a, b are both divisable by the monomial in y_1, y_2 spanning $\hat{A}_\lambda$. Consequently we can further assume $\lambda = 0$. All this now gives that $a \in K\{a_1^i a_2^j : i, j = 0, 1, 2, \cdots, \ell - 1\}$. Now $[x_1, a_2]$ commutes with y_2 and has degree $< \ell$ in x_2. Hence $[x_1, a_2]$ is a scalar and this scalar equals

$(\ell - 1)!(-1)^{\ell-2} = 1$. Similarly $[x_2, a_1] = 1$. Using the nilpotent action of ad x_1, ad x_2 which results we conclude that we may assume $a \in K \setminus \{0\}$, as required.

6.6. Retain the notation and hypotheses of 6.5.

COROLLARY. *The skew fields* Fract $U(\mathfrak{g})$ *and* $D_2(z)$ *are not isomorphic.*

Suppose in fact we had an isomorphism $\varphi :$ Fract $U(\mathfrak{g}) \xrightarrow{\sim} D_2(z)$. Observe that the specialization $U(\mathfrak{g})_{z=r} : r \in \mathbb{Z}$ is still an Ore domain. Consider say $\varphi(x_1)$ which takes the form $a^{-1}b : a, b \in D_2(z)$, $a \neq 0$. The set of integers for which $a = 0$ at the specialization $z = s$, is finite. A similar statement holds for all the finitely many generators of Fract $U(\mathfrak{g})$ and $D_2(z)$. It follows that we can find integers $r \neq 0$, s such that φ defines an isomorphism $\text{Fract}(U(\mathfrak{g})_{z=r}) \xrightarrow{\sim} D_2(z)_{z=s} \cong D_2$.

A similar argument shows that φ is defined over a finite extension $\mathbb{Q}(\xi_1, \xi_2, \cdots, \xi_n)$ of $\mathbb{Q}$. Now let A_0 (resp. B_0) the $\mathbb{Z}$ algebra with generators $x_i, y_i : i = 1, 2$, $\xi_j : j = 1, 2, \cdots, n$ and relations (1) with $z = r$ (resp. $z = 0$) supplemented by the relations defined by the minimal polynomials P_j of the ξ_j. Then we have an isomorphism $\varphi :$ Fract $A_0 \xrightarrow{\sim}$ Fract B_0. Set $A = A_0 \otimes_{\mathbb{Z}} \overline{\mathbb{F}}_\ell$, $B = B_0 \otimes_{\mathbb{Z}} \overline{\mathbb{F}}_\ell$ where ℓ does not divide r. Then A, B are again Ore domains and so further choosing ℓ so that the denominators of the images of the finitely many generators do not vanish we obtain an isomorphism $\varphi :$ Fract $A \xrightarrow{\sim}$ Fract B. Finally since $\overline{\mathbb{F}}_\ell$ is algebraically closed we have Fract $A = E$, Fract $B = D$. Yet E, D are not isomorphic because of 6.5. This contradiction establishes the corollary.

6.7. One could hope to generalize 6.6 for any completely solvable Lie algebra $\mathfrak{g}$. Let $P \in$ Spec $U(\mathfrak{g})$ and $E := E(P; \mathfrak{g})$ the set of non-zero ad $\mathfrak{g}$ eigenvectors of $U(\mathfrak{g})/P$. We remark that $U(\mathfrak{g})/P$ is an integral domain and the elements of $E(P, \mathfrak{g})$ commute ([52, Chap. 14]). In a series of papers J.C. McConnell ([51], and references therein) completely classified the algebras which can arise as localizations $E^{-1}(U(\mathfrak{g})/P)$. One can ask which of these McConnell algebras remain non-isomorphic when one passes to the full ring of quotients.

6.8. Two further relatively elementary uses of reduction modulo p are worthy of note. The first is the classification of the fixed ring $A_1(\mathbb{C})^G$ where G is a finite subgroup of Ant $A_1(\mathbb{C})$ by J. Alev, T.J. Hodges, J.-D. Veleg [1]. Here the technique of reduction mod p allows one to apply the classification of rational double points to distinguish these algebras. Second, the work of T.J. Stafford [64] showing that a projective left ideal P of $A_1(k) : k$ infinite, must be cyclic if one is to have an isomorphism $\text{End}_{A_1} P \xrightarrow{\sim} A_1$.

6.9.   We establish the promised version of Lie's theorem over any algebraically closed field k of arbitrary characteristic. Let V be a $\mathfrak{h}$ module and fix $\lambda \in (\mathfrak{h}, \mathfrak{h})^{\perp}$. Then λ defines a one-dimension representation $\rho_\lambda : x \mapsto \lambda(x)1$ of $\mathfrak{h}$. Let $\ker \rho_\lambda$ denote the kernel of ρ_λ viewed as a representation of $U(\mathfrak{h})$. For each $i \in \mathbb{N}$, set

$$V^{\lambda,i} = (\ker \rho_\lambda)^i V.$$

Obviously the $V^{\lambda,i}$ form an increasing family of $\mathfrak{h}$ submodules of V and we set

$$V^\lambda = \bigcup_{i \in \mathbb{N}} V^{\lambda,i}.$$

LEMMA.   Let $\mathfrak{g}$ be a Lie algebra and $\mathfrak{h}$ an ideal of $\mathfrak{g}$. Let V be a $U(\mathfrak{g})$ module. Fix $\lambda \in \mathfrak{h}^*$. Then for each $i \in \mathbb{N}$ one has $\mathfrak{g}V^{\lambda,i} \subset V^{\lambda,i+1}$. In particular V^λ is a $\mathfrak{g}$ submodule of V.

The proof is by induction on i. Since $V^{\lambda,0} = 0$ it holds trivially for $i = 0$. Now for all $x \in \mathfrak{g}$, $y \in \mathfrak{h}$ one has

$$(x - \lambda(x))yV^{\lambda,i} \subset [x,y]V^{\lambda,i} + y(x - \lambda(x))V^{\lambda,i}$$
$$\subset V^{\lambda,i} + yV^{\lambda,i-1} \subset V^{\lambda,i}$$

by the induction hypothesis. Thus $(\ker \rho_\lambda)^{i+1}yV^{\lambda,i} = 0$ as required.

6.10.   Let $\mathfrak{g}$ be a finite dimensional solvable Lie algebra over an algebraically closed field k of any characteristic. Let V be a finite dimensional $\mathfrak{g}$ module. The usual induction argument ([14, 1.3.12]) applied to 6.9 gives

COROLLARY.   There exists $\lambda \in (\mathfrak{g}, \mathfrak{g})^{\perp}$ such that $V^\lambda \neq 0$.

§7. Equivalences of categories

This section concerns a semisimple Lie algebra $\mathfrak{g}$ over an algebraically closed field k of characteristic zero. Fix a triangular decomposition $\mathfrak{g} = \mathfrak{n} \oplus \mathfrak{h} \oplus \mathfrak{n}^-$ and let ρ denote the corresponding half sum of the positive roots. For each $\lambda \in \mathfrak{h}^*$ let $M(\lambda)$ denote the Verma module ([14, Sect. 7.1]) with highest weight $\lambda - \rho$ and $L(\lambda)$ its unique simple quotient. Set $\mathfrak{b} = \mathfrak{h} \oplus \mathfrak{n}$. Let W denote the Weyl group for the pair $\mathfrak{g}$, $\mathfrak{h}$.

7.1.   Let $\mathcal{O}$ denote the category of all finitely generated $U(\mathfrak{g})$ modules M.

(i) $\mathfrak{b}$ acts locally finitely on M, that is $\dim U(\mathfrak{b})m < \infty$, $\forall m \in M$.
(ii) M is a direct sum of its finite dimensional weight spaces $M_\lambda : \lambda \in \mathfrak{h}^*$.

It is easy to show that dim $M_\lambda < \infty$, that is M admits a formal character ch(M), for example see ([14, 7.5]). Every simple subquotient of $M \in \mathrm{Ob}\,\mathcal{O}$ is some $L(\lambda) : \lambda \in \mathfrak{h}^*$. Again Ann $M(\lambda) \in \max Z(\mathfrak{g})$ and by the Harish-Chandra theorem ([14, Sect. 7.4]) the map $\lambda \mapsto$ Ann $M(\lambda)$ factors to isomorphism of $\mathfrak{h}^*/W$ onto max $Z(\mathfrak{g})$. It easily follows that only finitely many distinct $L(\lambda)$ can occur as subquotients of M and then that M has finite length.

7.2. The map $x \mapsto -x$ on $\mathfrak{g}$ extends to an antiautomorphism $a \mapsto \check{a}$ of $U(\mathfrak{g})$. Via this antiautomorphism we may view any $U(\mathfrak{g})$ bimodule V as a $U(\mathfrak{g}) \otimes U(\mathfrak{g})$ module and hence as a $U(\mathfrak{g} \times \mathfrak{g})$ module. Let $\mathfrak{k}$ denote the diagonal copy of $\mathfrak{g}$ in $\mathfrak{g} \times \mathfrak{g}$. Note the action of $\mathfrak{k}$ on V identifies with the adjoint action of $\mathfrak{g}$ defined in 2.9. Let $\mathcal{H}$ denote the category of all $\mathfrak{g} \times \mathfrak{g}$ modules V such that

(i) V is a direct sum of finite dimensional simple $U(\mathfrak{k})$ modules having finite multiplicity.

(ii) $\mathrm{Ann}_{Z(\mathfrak{g}) \otimes Z(\mathfrak{g})} V$ has finite codimension.

It is less obvious that the objects of $\mathcal{H}$ have finite length but this can be established by an equivalence of categories theorem ([6], [18]). First by (ii) we can assume without loss of generality that the radical of $\{a \in Z(\mathfrak{g}) \mid V a = 0\}$ is the maximal ideal Ann $M(\lambda)$. We let $\mathcal{H}_\lambda$ denote the subcategory of all $V \in \mathrm{Ob}\,\mathcal{H}$ with this property.

Call $\lambda \in \mathfrak{h}^*$ dominant if $2(\lambda, \alpha)/(\alpha, \alpha) \notin \{-1, -2, \cdots\}$ for every positive root α. Given $\mu \in \mathfrak{h}^*$ we may choose $\lambda \in W\mu$ dominant. Assume $\lambda \in \mathfrak{h}^*$ dominant. One easily checks via the action of $Z(\mathfrak{g})$ that $M(\lambda)$ is projective in $\mathcal{O}$. This provides an exact functor $T : M \mapsto F(M(\lambda), M)$ from $\mathcal{O}$ to $\mathcal{H}_\lambda$. The functor $T' : V \mapsto V \otimes_{U(\mathfrak{g})} M(\lambda)$ is a left adjoint for T. Call $\lambda \in \mathfrak{h}^*$ regular if $(\lambda, \alpha) \neq 0$, for every non-zero root α.

THEOREM.

(i) $V \xrightarrow{\sim} T(T'(V))$, $\forall V \in \mathrm{Ob}\,\mathcal{H}_\lambda$
(ii) If λ is regular T is an equivalence of categories.

7.3. We apply 7.2 to recover an embedding result for prime Dixmier algebras in the semisimple case. This was implicit in [43] in a slightly different form. A slightly weaker version was noted independently by I. Pranata [58].

Observe that by definition (2.8), a Dixmier algebra D viewed as a $U(\mathfrak{g})$ bimodule belongs to $\mathcal{H}_\lambda$ for some λ dominant. Set $M = T'(D) = D \otimes_{U(\mathfrak{g})} M(\lambda)$.

COROLLARY. *Assume D to be a prime ring. Then the map $D \to \mathrm{End}_k M$ defined by left multiplication is injective.*

Set $K = \ker(D \to \mathrm{End}_k\, M)$. By (i) we have an isomorphism $D \xrightarrow{\sim} T(M)$ of $U(\mathfrak{g})$ bimodules. Thus if J denotes the annihilator of D considered as a left $U(\mathfrak{g})$ module the composed map $U(\mathfrak{g})/J \hookrightarrow D \to D/K$ is an embedding.

Since D has finite length as a $U(\mathfrak{g})$ bimodule, 7.2(i) implies that it is finitely generated as a left $U(\mathfrak{g})$ module. In particular D is a noetherian ring. Suppose D is prime and $K \neq 0$. Then by Goldie's theorem there exists $a \in K$ regular in D. Let d denote Gelfand-Kirillov dimension with respect to $U(\mathfrak{g})$. Then

$$d(U(\mathfrak{g})/J) \leq d(D/K) \leq d(D/Da) < d(D) \leq d(U(\mathfrak{g})/J).$$

This contradiction proves that $K = 0$.

REMARKS.   This identifies D as a subring of $F(M)$. A basic question is whether M is quasi semisimple, that is if $d(M/\mathrm{Soc}\ M) < d(M)$. Then as in 5.8 we obtain an embedding $F(M) \hookrightarrow F(\mathrm{Soc}\ M)$. Now if we write $\mathrm{Soc}\ M = \oplus L_i$, then

$$F(\mathrm{Soc}\ M) \cong \oplus F(L_i, L_j)$$

where the ring structure is given by composition of endomorphisms. By ([43, 2.9]) each $F(L_i, L_j)$ is canonically determined by Fract $F(L_i, L_j)$ and by ([31, III, 4.13]) is quasi-semisimple. Further detailed information on the $F(L_i, L_j)$ obtains from ([35, 2.3 and Sect. 4]).

7.4.   Fix $\lambda \in \mathfrak{h}^*$ and set $\Lambda = \lambda + P(R)$ where $P(R)$ denotes the lattice of weights. Let $\mathcal{O}_\Lambda$ denote the subcategory of $\mathcal{O}$ of modules whose weights lie in Λ. Let Λ^+ (resp. Λ^{++}) denote the dominant (resp. and regular) elements of Λ.

PROPOSITION.   Take $M_1, M_2 \in \mathrm{Ob}\,\mathcal{O}_\Lambda$. Then $A(M_1, M_2)$ is finitely generated as a $U(\mathfrak{g})$ bimodule.

Take $\lambda_1 \in \Lambda^{++}$. As noted in ([37, 3.1.1]) it is a consequence of 7.2 that any $M_1 \in \mathrm{Ob}\,\mathcal{O}_\Lambda$ can be expressed as an image of $M(\lambda_1) \otimes E_1$ for some finite dimensional module E_1. Then $A(M_1, M_2) \hookrightarrow A(E_1 \otimes M(\lambda_1), M_2) \xrightarrow{\sim} E_1^* \otimes A(M(\lambda_1), M_2)$, where the last step is Stafford's isomorphism (cf. [36, 3.2]). Again since objects in $\mathcal{O}_\Lambda$ have finite length we can assume M_2 is simple without loss of generality, say $M_2 = L(\lambda_2)$. Note that $\lambda_1 - \lambda_2 \in P(R)$.

Set $A(M(\lambda_1), M(\lambda_1)) = A(\lambda_1)$ and consider $A(M(\lambda_1), L(\lambda_2))$ as a right $A(\lambda_1)$ module.   We   claim   that   it   equals   its   submodule $F(M(\lambda_1), L(\lambda_2))A(\lambda_1)$.

Take $\nu \in P(R)^+$. Then $\lambda_1 + \nu \in \Lambda^{++}$. By ([43, 5.3]) the map $\theta_{\lambda_1+\nu}^{\lambda_1} \in A(M(\lambda_1+\nu), M(\lambda_1))^{\mathfrak{n}^-}$ which takes the highest weight vector of $M(\lambda_1+\nu)$

to the highest weight vector of $M(\lambda_1)$, actually lies in $F(M(\lambda_1+\nu), M(\lambda_1))$. Consequently

$$F(M(\lambda_1 + \nu), L(\lambda_2)) = F(M(\lambda_1), L(\lambda_2))F(M(\lambda_1 + \nu), M(\lambda_1))$$

since the right hand side is a non-zero $U(\mathfrak{g})$ bisubmodule of the left hand side which is simple by 7.2. Recalling ([43, 5.5]) we obtain

$$\varinjlim F(M(\lambda_1 + \nu), L(\lambda_2))$$
$$= F(M(\lambda_1), L(\lambda_2)) \varinjlim F(M(\lambda_1 + \nu), M(\lambda_1))$$
$$= F(M(\lambda_1), L(\lambda_2))A(\lambda_1).$$

It remains to show that the left hand side equals $A(M(\lambda_1), L(\lambda_2))$. This follows exactly as in ([43, 5.5]); but let us make a simplification in that argument.

Recall the polynomial subring P of $A(\lambda_1)$ described in ([36, 3.3]). One easily checks using Taylor's lemma and the behaviour of P under the adjoint section of $\mathfrak{n}^-$ (see [36, 3.3]) that right multiplication gives a surjection

$$A(M(\lambda_1), L(\lambda_2))^{\mathfrak{n}^-} \otimes P \twoheadrightarrow A(M(\lambda_1)L(\lambda_2)).$$

Now identify $A(M(\lambda_1), L(\lambda_2))^{\mathfrak{n}^-}$ with $L(\lambda_2)$. It is then enough to show that any weight vector $a \in L(\lambda_2)$ viewed as element of $A(M(\lambda_1+\nu), L(\lambda_2))^{\mathfrak{n}^-}$ actually lies in $F(M(\lambda_1 + \nu), L(\lambda_2))^{\mathfrak{n}^-}$ for $\nu \in P(R)^+$ sufficiently large. This follows exactly as in ([43, 5.3]).

Finally the simplicity of $F(M(\lambda_1), L(\lambda_2))$ implies that we can find a finite dimensional $\mathfrak{k}$ submodule E over which is generated as a left $U(\mathfrak{g})$ module. On the other hand via left multiplication we can identify $U(\mathfrak{n}^-)$ with a subring of $A(\lambda_1)$ and it is an easily proved (and well-known) fact that $A(\lambda_1) = PU(\mathfrak{n}^-)$. Consequently $A(M(\lambda_1), L(\lambda_2)) = U(\mathfrak{g})EPU(\mathfrak{n}^-)$. Now recall that P is itself a $U(\mathfrak{k})$ submodule which lies in the $\mathcal{O}$ category. Thus EP, which identifies with a quotient of $E \otimes P$, again lies in the $\mathcal{O}$ category and so has finite length. In particular EP is a finitely generated $U(\mathfrak{k})$ module. Thus we can write $EP = U(\mathfrak{k})H : \dim H < \infty$ and from the above H generates $A(M(\lambda_1), L(\lambda_2))$ as a $U(\mathfrak{g})$ bimodule.

7.5.   One can show that $A(M_1, M_2) : M_1, M_2 \in \mathrm{Ob}\,\mathcal{O}_\Lambda$ has even finite length as $U(\mathfrak{g})$ bimodule; but this will need a more sophisticated theory. First let us note some general well-known but nevertheless remarkable facts about flat overrings.

Let U be an Ore domain and F its field of fractions. Let A be an overring of U contained in F which is flat as a right U module. Let R be any ring.

LEMMA.

(i) *The map $a \otimes b \overset{\mu}{\mapsto} ab$ of $A \otimes_U A$ onto A is bijective.*

(ii) *Let M be a $U - R$ bimodule of finite length. Then $A \otimes_U M$ is an $A - R$ module of finite length..*

(iii) *For any left A module N the multiplication map is an isomorphism of $A \otimes_U N$ onto N.*

(i)   Let $\mu' : A \otimes_U F \to F$ be the multiplication map. It is clear that $\mu = \mu'i$, where i is the map $A \otimes_U A \to A \otimes_U F$ obtained from the embedding $A \hookrightarrow F$. Since A is right U flat, i is injective. Again let $\mu'' : F \otimes_U F \to F$ be the multiplication map. Then $\mu' = \mu''j$, where j is the map $A \otimes_U F \to F \otimes_U F$ obtained from the embedding $A \hookrightarrow F$. Since F is left U flat, j is bijective. It remains to show that μ'' is injective. Since the right U module $U \setminus F$ is torsion we have $U \setminus F \otimes_U F = 0$. It follows that the injection $U \hookrightarrow F$ leads to a *surjection* $U \otimes_U F \to F \otimes_U F$. Thus every element of $F \otimes_U F$ can be written in the form $1 \otimes f : f \in F$ and consequently μ'' is injective.

(ii)   We can assume M simple without loss of generality. Let $\varphi : m \mapsto 1 \otimes m$ be the canonical map of M in $A \otimes_U M$. Obviously $A\varphi(M) = A \otimes_U M$, so we can assume that $\varphi(M) \neq 0$. Moreover it will be enough to show that for any non-zero $A - R$ submodule N of $A \otimes_U M$ one has $N \cap \varphi(M) \neq 0$. If not the composed map $N \hookrightarrow A \otimes_U M \longrightarrow A \otimes_U M/\varphi(M)$ of U modules is an injection. Since A is right U flat we obtain an injection

$$A \otimes_U N \hookrightarrow A \otimes_U (A \otimes_U M/\varphi(M)).$$

Yet by (i) the map $A \otimes_U \varphi(M) \longrightarrow A \otimes_U A \otimes_U M$ is surjective and so the right hand side of (∗) is zero. From the surjection $a \otimes n \mapsto an$ of $A \otimes_U N$ onto N we conclude that $N = 0$, as required.

(iii)   One has $N = A \otimes_A N = (A \otimes_U A) \otimes_A N = A \otimes_U (A \otimes_A N) = A \otimes_U N$.

7.6.   To establish the finite length property discussed in 7.5 we use the Beilinson-Bernstein equivalence of categories theorem [3] in the form given by Hodges-Smith [25]. Take as before $\lambda_1, \lambda_2 \in \Lambda^{++}$. Set $U = U(\mathfrak{g})/\operatorname{Ann} M(\lambda_2)$, $A = A(\lambda_2)$. By ([43, 5.5]), A is left U flat. By ([14, 8.4.4 and 5.3]), U is an Ore domain. Again the analysis in ([43, 5.3 - 5.5]) was actually a modification of the proof [13] that A embeds in $F := \operatorname{Fract} U$. Then by 7.5(ii) the multiplication map gives an isomorphism.

$$F(M(\lambda_1), L(\lambda_2)) \otimes_U A \overset{\sim}{\longrightarrow} A(M(\lambda_1), L(\lambda_2))$$

which is furthermore a simple $U(\mathfrak{g}) - A$ bimodule.

Now recall that the Weyl group W acts by automorphisms in $U(\mathfrak{g})$ leaving stable any two-sided ideal. We deduce from this an action of W on U and hence on F. Then for each $w \in W$ let A_w denote the image of A under $w \in \operatorname{Aut} F$. It is clear that A_w is again left U flat. Now recall that A admits a polynomial subring P. It turns out that the action of W on F is rather easily described as follows. Recall that P is $\mathfrak{k}$ stable and hence Fract P is stable under the corresponding algebraic subgroup K of $G \times G$. Representatives of W may be chosen ([14, 1.10.19]) from K and so Fract P is W stable. Again this action of W must be compatible with the $\mathfrak{t}$ module structure of P (where $\mathfrak{t}$ is the image of $\mathfrak{h}$ in $\mathfrak{k}$ under the diagonal map). Now recall ([36, 3.3]) that P is generated by $\mathfrak{t}$ weight vectors p_γ of weight $\gamma \in R^+$ as a polynomial ring, moreover as a $\mathfrak{t}$ module it identifies with the $\mathcal{O}$ dual $\delta M(\rho)$ of the Verma module with *lowest* weight O (that is taking $\mathfrak{b} = \mathfrak{n}^- \oplus \mathfrak{h}$). From this it is quite easy to conclude that for each simple reflection s_α one has

$$
s_\alpha p_\gamma = \begin{cases} p_{s_\alpha \gamma} : \gamma \neq \alpha \ , \\[2ex] p_\alpha^{-1} \ \ : \gamma = \alpha. \end{cases}
$$

Now for each $w \in W$, set $S(w) = \{\alpha \in R^+ \,|\, w\alpha \in R^-\}$ and

$$
f_w = \prod_{\gamma \in S(w^{-1})} p_\gamma.
$$

Then $wP \subset P[f_w^{-1}]$. Finally recall ([36, 3.3]) that A identifies with the subring of $\operatorname{End}_k P$ of differential operators ([52, Chap. 15]) on P. Thus the action of W on P induces an action of W on $A \otimes_P$ Fract P and in particular

$$
A_w \subset A[f_w^{-1}].
$$

Moreover the right hand side is just the subring B of Fract A generated by A, A_w. We have a commutative diagram

$$
\begin{array}{ccc}
A \otimes_U A_w & \xrightarrow{\;i\;} & B \otimes_U A_w \\[2ex]
{\scriptstyle \mu} \searrow & & \downarrow {\scriptstyle \mu'} \\[2ex]
& B &
\end{array}
$$

where i obtains from the embedding $A \hookrightarrow B$ and is injective since A_w is left U flat and μ, μ' are multiplication. By (iii) μ' is bijective so μ is injective.

Finally from the above description of B one has $B = AA_w$ so μ is surjective. We have the

LEMMA.   *Take $w \in W$ and define $f_w \in P$, $A_w \subset$ Fract U as above. Then the multiplication map is an isomorphism of $A \otimes_U A_w$ onto $A[f_w^{-1}]$.*

REMARK.   Since A is also left U flat, multiplication gives $A_w \otimes_U A \xrightarrow{\sim} A [f_w^{-1}]$.

7.7.   The really deep fact which obtains from the Hodges-Smith interpretation of Beilinson-Bernstein (and which for the moment has apparently no simple ring theoretic proof) is the following

THEOREM.   *Let L be a simple right U module. Then there exists $w \in W$ such that $L \otimes_U A_w \neq 0$. That is $\oplus_{w \in W} A_w$ is faithfully left U flat.*

7.8.   The above results carry over with respect to right action, though here we have to be careful. Thus A is *not* right U flat. Rather we must choose $\lambda_3 \in W\lambda_2 \cap \Lambda$ antidominant and set $A' = A(\lambda_3)$. Then A' is a right $U' := U(\mathfrak{g})/$ Ann $M(\lambda_2)$ flat ([43, 5.6]). Moreover defining $A'_y : y \in W$ as in 7.6 we further obtain that the overring

$$\bigoplus_{y \in W} A'_y$$

is faithfully right U' flat. In the special case $\lambda_2 = \rho$ we note that $F(M(\rho), M(\rho))$ coincides with the ring of global differential operators on the flag variety G/B. The $A_w : w \in W$ are local sections of the sheaf $\mathcal{D}$ of differential operators on G/B corresponding to the open Bruhat cell and W translates of it. In this context one translates from left to right $\mathcal{D}$ modules by tensoring with the sheaf of differentials on G/B of top degree. The connection here can be seen by noting that $\lambda_2 - \lambda_3 = 2\rho$ which is just the weight of the top exterior product

$$\Lambda_{\alpha \in R^+} x_\alpha \quad : \quad x_\alpha \in \mathfrak{n} \text{ of weight } \alpha.$$

Retain the notation of 7.6 and set $F_{12} = F(M(\lambda_1)L(\lambda_2))$. Recall 5.7. Observe that $A'_y \otimes A_w^{opp}$ is a Weyl algebra.

LEMMA.   *For each $y \in W$, the Weyl algebra module $M := A'_y \otimes_{U'} F_{12} \otimes_U A_w$ is simple and holonomic.*

Simplicity follows from 7.5(ii) and the remarks in 7.6. To prove that M is holonomic, we must show that $d(M) \leq |R|$ for M considered as a $A'_y - A_w$ bimodule. Let $f \in F_{12}$ be a non-zero highest $\mathfrak{k}$ weight vector of weight ν. Let L denote the kernel of the surjective map $A'_y \otimes_k A_w \longrightarrow M$. Then $x_\alpha \otimes 1 + 1 \otimes \alpha_\alpha \in L$, $\forall\, \alpha \in R^+$ and since dim $U(\mathfrak{k})f < \infty$ there exists $\ell \in \mathbb{N}$ such that $(x_\alpha \otimes 1 + 1 \otimes x_\alpha)^\ell \in L$, $\forall\, \alpha \in -R^+$. Again $h \otimes 1 + 1 \otimes h - \nu(h)(1 \otimes 1) \in L$, $\forall\, h \in \mathfrak{h}$. Then with respect to any (finite) filtration $\mathcal{F}$ of $A'_y \otimes A_w$ it follows that $\mathfrak{k} \subset \sqrt{\mathrm{gr}\ L}$. Since A_w, U have the same ring of fractions this forces $d(M) \leq d(A'_y) = |R|$ as required.

REMARK.   One may recognize that M is the space of sections on an appropriate translate of the open Bruhat cell of $G/B \times G/B$ for the Harish-Chandra sheaf $\mathcal{M}$ for $\mathcal{D}' \otimes \mathcal{D}$ obtained from the Harish-Chandra module $F(M(\lambda_1), L(\lambda_2))$ by localization. In this sense the assertion of the lemma is well-known.

7.9.   Retain the notation of 7.6–7.8 and set $A_{12} = A(M(\lambda_1), L(\lambda_2))$. We wish to show for all $y, w \in W$ that $A'_y \otimes_{U'} \otimes A_{12} \otimes_U A_w$ has finite length as an $A'_y - A_w$ bimodule. For this we need to discuss how a holonomic module behaves under localization. Here we use a filtration argument which goes back to I.N. Bernstein [5] and is of interest in its own right.

Give the Weyl algebra $A = k[x_i, \frac{\partial}{\partial x_j} : i, j = 1, 2, \cdots, n]$ a filtration $\mathcal{F}$ by taking x_i, $\frac{\partial}{\partial x_i}$ to have degree 1. Let M be a finitely generated A module. Then M admits a finite filtration $\mathcal{F}$ as constructed in (4.2). Let $f \in k[x_i : i = 1, 2, \cdots, n]$ be any non-zero polynomial. Set $s = \deg f$. One checks that

$$\mathcal{F}^\ell M[f^{-1}] = \{ f^{-\ell} m : m \in \mathcal{F}^{(s+1)\ell}(M) \}$$

is a filtration (not-necessarily finite) of $M[f^{-1}]$. The following result is due to Bernstein [5].

THEOREM.   (char $k = 0$). Suppose $d(M) \leq n$. Then M has finite length as an A module.

We can assume $M \neq 0$. Then by (5.5), $d(M) = n$ and furthermore $\dim_k \mathcal{F}^\ell(M)$ is a polynomial for ℓ sufficiently large, hence of the form $e\ell^n/n! + O(\ell^{n-1})$ with $e \in \mathbb{N}^+$. It follows that $\dim_k \mathcal{F}^\ell(M[f^{-1}])$ is bounded by a polynomial with leading term $e(s+1)^n \ell^n/n!$. Now consider an increasing sequence $M_0 \subset M_1 \subset M_2 \cdots$ of finitely generated A submodules of $M[f^{-1}]$. Since $d(M_{i+1}/M_i) \geq n$ if $M_{i+1}/M_i \neq 0$, we conclude that the induced filtration $\mathcal{F}$ on this quotient satisfies $\dim_k \mathcal{F}^\ell(M_i/M_{i+1}) = e_i \ell^n/n! + O(\ell^{n-1})$ with $e_i \in \mathbb{N}^+$. By additivity on exact sequences $\sum e_i \leq e(s+1)^n$ and so the sequence must become stationary after $\leq e(s+1)^n$ steps. Similarly any decreasing sequence is stationary. Hence $M[f^{-1}]$ has finite length.

7.10.   We now conclude with our main result.

THEOREM.   *Take $M_1, M_2 \in \mathrm{Ob}\,\mathcal{O}_\Lambda$. Then $A(M_1, M_2)$ has finite length as a $U(\mathfrak{g})$ bimodule.*

By 7.7 it is enough to show for all $y, w \in W$ that $A'_y \otimes_{U'} A_{12} \otimes_U A_w$ has finite length as an $A'_y - A_w$ bimodule. Set $f = w(f_w)$. By 7.6 the multiplication maps give isomorphisms $A \otimes_U A_w \xrightarrow{\sim} A_w[f^{-1}] \xleftarrow{\sim} A_w \otimes_U A$. Then by 7.8 the above module is just $A'_y \otimes_{U'} A_{12} \otimes_U A_w = A'_y \otimes_{U'} F_{12} \otimes_U A \otimes_U A_w = M[f^{-1}]$ as a $A'_y - A_w$ bimodule. Since M is holonomic the result follows from 7.9.

REMARK.   The reader might be tempted to think that this also follows from sheaf theoretic mumbo-jumbo. Yet although $A'_y \otimes_{U'} A_{12} \otimes_U A_w$ looks like the sections of a Harish-Chandra sheaf, in fact we obtain at each step the sections corresponding to a smaller open set. Moreover the result is almost certainly false if we replace F_{12} by an arbitrary simple bimodule, since as shown by T.J. Stafford ([63, Cor. 1.2]) localization in the sense of 7.9 can destroy the finite length property.

§8. Homological algebra

8.1.   A large number of results in enveloping algebras, especially character formulae are obtained using homological algebra. In particular the structure of semisimple Lie algebras is sufficiently rich to often admit closed formulae.

8.2.   We start with a result which can be read off from Hochschild-Serre [24] or checked directly. Let $\mathfrak{p}$ be a Lie algebra (not necessarily finite dimensional) $\mathfrak{n} \supset \mathfrak{m}$ ideals of $\mathfrak{p}$ with dim $\mathfrak{n}/\mathfrak{m} = 1$. Let F be a $\mathfrak{p}$ module.

LEMMA.   *For each $i \in \mathbb{N}$ one has an exact sequence*

$$0 \to H_0(\mathfrak{n}/\mathfrak{m}, H_i(\mathfrak{m}, F)) \to H_i(\mathfrak{n}, F) \to H_1(\mathfrak{n}/\mathfrak{m}, H_{i-1}(\mathfrak{m}, F)) \to 0$$

of $\mathfrak{p}$ modules.

8.3.   We apply 8.2 to a Kac-Moody Lie algebra $\mathfrak{g}$. Such a Lie algebra admits a triangular decomposition $\mathfrak{g} = \mathfrak{n}^+ \oplus \mathfrak{h} \oplus \mathfrak{n}$ (note our change of convention). Fix a simple root α. Then $s_\alpha = k\{x_a, x_\alpha, h_\alpha\}$ is an $\mathfrak{sl}(2)$ subalgebra. Set $\mathfrak{r}_\alpha = \mathfrak{h} + s_\alpha$. Then $\mathfrak{p} = \mathfrak{n} \oplus \mathfrak{h} \oplus kx_\alpha$ is a (parabolic) subalgebra with nilradical $\mathfrak{m} \subset \mathfrak{n}$ complemented by $\mathfrak{r}_\alpha$ in $\mathfrak{p}$.

Now assume that F in 8.2 is a direct sum of finite dimensional $\mathfrak{r}_\alpha$ modules in which $\mathfrak{h}$ acts semisimply. This property carries over to $H_i(\mathfrak{m}, F)$. Now for any $\mathfrak{h}$ module M and any $\mu \in \mathfrak{h}^*$, set $M_\mu = \{m \in M \mid Hm = \mu(H)m, \ \forall\, h \in \mathfrak{h}\}$. Let s_α denote the reflection corresponding to α. Set $\alpha^\vee = 2\alpha/(\alpha, \alpha)$.

LEMMA. *For each $\nu \in \mathfrak{h}^*$ such that $(\alpha, \nu) \geq 0$ one an isomorphism*

$$H_i(\mathfrak{n}, F)_\nu \xrightarrow{\sim} H_{i+1}(\mathfrak{n}, F)_{s_\alpha \nu - \alpha}.$$

Moreover both sides are zero unless $(\alpha^\vee, \nu) \in \mathbb{N}$.

Let E be a simple $\mathfrak{r}_\alpha$ direct summand of $H_i(\mathfrak{m}, F)$. Then E is completely determined by its highest weight μ. Obviously $E^{x-\alpha} \cong k_{s_\alpha \mu}$ (where the right hand side is the one dimension $\mathfrak{h}$ module spanned by a vector of weight $s_\alpha \mu$) and so $H_1(\mathfrak{n}/\mathfrak{m}, E) \cong H^0(kx_{-\alpha}, E) \cong kx_{-\alpha} \otimes E^{x-\alpha} \cong k_{s_\alpha \mu - \alpha}$. Again $E/x_\alpha E \cong k_\mu$ and so $H_0(\mathfrak{n}/\mathfrak{m}, E) \cong k_\mu$. Then the assertion of the lemma follows from 8.2.

8.4. Call a $\mathfrak{g}$ module L integrable if it has the property enjoyed by F in 8.3 for every simple root α. We can almost completely determine ch $H_*(\mathfrak{n}, L)$ by 8.3; but there is one snag when dim $\mathfrak{n}$ is not finite. To understand this choose $\delta \in \mathfrak{h}^*$ such that $(\alpha^\vee, \delta) = 1$ for every simple root and define the translated action of the Weyl group W on $\mathfrak{h}^*$ through $w \cdot \mu = w(\mu + \delta) - \delta$. Observe that $s_\alpha \cdot \mu = s_{\alpha \mu - \alpha}$. Our assumption on L implies that every weight of L and hence of $H_i(\mathfrak{n}, L)$ is integral. Then by 8.3, we find that $s_\alpha . \nu$ is a weight of $H_{i+1}(\mathfrak{n}, L)$. Then by induction for each $w \in W$, we conclude that $w \cdot \nu$ is a weight of $H_{i+\ell(w)}(\mathfrak{n}, L)$ where $\ell(w)$ denotes the reduced length of w. If dim $\mathfrak{n} < \infty$, there exists a unique (longest) element $w_0 \in W$ such that $\ell(w_0) = \dim \mathfrak{n}$. Since $H_j(\mathfrak{n}, L) = 0$ for $j > \dim \mathfrak{n}$, we conclude that $i = 0$ in the above. Moreover $H_0(\mathfrak{n}, L) = L/\mathfrak{n}L$ is just the span of the highest weight spaces of L which is one dimensional if L is simple and characterizes L in general.

We can recover a similar result in general if we assume that $\mathfrak{g}$ is symmetrizable. Assume further that L is a simple module of highest weight μ. Then every weight of $H_i(\mathfrak{n}, L)$ is a weight ν of L plus a weight of $\Lambda^i \mathfrak{n}$, hence of the form $\nu := \mu - \sum k_\alpha \alpha : k_\alpha \in \mathbb{N}$ where the sum is over the simple roots. Identify $H_i(\mathfrak{n}, L)_\nu$ with $H^i(\mathfrak{n}^+, L)_\nu = \operatorname{Ext}^i(M(\nu), L)$ via the contravariant form on L (which is non-degenerate) where $M(\nu)$ denotes the Verma module of highest weight ν. Since $\operatorname{Ext}^i(M(\nu), L) \neq 0$ only if the Casimir invariant acts by the same scalar on $M(\nu)$ and on L, we further conclude that $(\nu + \delta, \nu + \delta) = (\mu + \delta, \mu + \delta)$. An easy though crucial classical combinatorial result of V.G. Kac ([44, 10.3]) then asserts that ν can only be dominant if $\nu = \mu$. This forces $i = 0$. We obtain the

THEOREM. *Let $\mathfrak{g}$ be a symmetrizable Kac-Moody Lie algebra and $L(\mu)$*

is simple integrable highest weight $\mathfrak{g}$ module with highest weight μ. Then

$$
\dim\ H_i(\mathfrak{n}, L)_\nu = \begin{cases} 1 : \nu = w \cdot \mu \text{ and } \ell(w) = 1 \\ \\ 0 : \text{ otherwise.} \end{cases}
$$

REMARK.   This recovers the famous Bott-Kostant theorem. The possibility of giving such an easy proof was first noted in ([34, 2.19]). Further applications are given in [34, II].

8.5.   One can go on to give a resolution of $L(\mu)$ by the free $\mathfrak{n}^-$ modules $M(w \cdot \mu)$. A standard argument ([14, 7.8.14]) gives an embedding $M(s_\alpha w \cdot \mu) \hookrightarrow M(w \cdot \mu)$ wherever α is a simple root satisfying $\ell(s_\alpha w) > \ell(w)$. A combinatorial property of the Bruhat order ([14, 7.8.14]) then gives a complex

$$
\to \oplus M(w \cdot \mu) \to \cdots \cdots \to M(\mu) \to L(\mu).
$$

COROLLARY.   *The complex above is exact.*

This is an easy induction argument (c.f. [18, Sect. 2]) based on the fact that for $M, N \in \mathrm{Ob}\,\mathcal{O}$ ([18, 1.10]) with N a submodule of M, *any* $\mathfrak{h}$ module isomorphism $H_0(\mathfrak{n}, N) \cong H_0(\mathfrak{n}, M)$ implies $N = M$. For more details see ([18, 1.10, Sect. 2]).

REMARK.   As pointed out by O. Mathieu (unpublished) this recovers the Gabber-Kac [19, 9.11] result describing the defining relations of a symmetrizable Kac-Moody Lie algebra. Another example of recovering resolutions from homology occurs in [59].

8.6.   There is an amusing application of 8.5 which seems to have gone unnoticed. It was obtained in discussions with J. Horvath. First let x_α denote the element of a Chevalley basis corresponding to the root α and let $\sigma : x_\alpha \longrightarrow x_{-\alpha}$ denote the corresponding Chevalley antiautomorphism. One has $\sigma \mid_{\mathfrak{h}} = Id$. Given $M \in \mathrm{Ob}\,\mathcal{O}$ give M^* a $\mathfrak{g}$ module structure through σ and let $\delta(M)$ denote the largest submodule of M^* on which $\mathfrak{h}$ acts locally finitely. An easy exercise shows that $\dim\ M_\mu = \dim\ \delta(M)_\mu$ for all $\mu \in \mathfrak{h}^*$. This further implies that $\delta M \in \mathrm{Ob}\,\mathcal{O}$ and that the contravariant functor δ on $\mathcal{O}$ is exact and involutory. Via the non-degenerate contravariant form on any $L \in \mathrm{Ob}\,\mathcal{O}$ simple one obtains $L \cong \delta L$.

Now let $\mathfrak{m}$ be *any* $\mathfrak{h}$ stable subalgebra of $\mathfrak{n}$. The standard isomorphism relating homology and cohomology gives an isomorphism of $\mathfrak{h}$ modules

(∗) $H_i(\mathfrak{m}, M)^* \xrightarrow{\sim} H^i(\sigma(\mathfrak{m}), M^*)$.

Given $M \in \mathrm{Ob}\,\mathcal{O}$ we may replace M^* by its submodule δM in the right hand side of $(*)$.

Now if M is a Verma module, hence $\mathfrak{n}$ free and a fortiori $\mathfrak{m}$ free, we obtain that the left hand side of $(*)$ vanishes for $i > 0$. We compute the right hand side of $(*)$ using 2.6, taking account of a possible multiplicity in the root spaces. By weight space decomposition, $\mathfrak{n}^+$ and hence $\sigma(\mathfrak{m})$ acts locally nilpotently on δM. Now in the standard identification of $M(\nu) := U(\mathfrak{g}) \otimes_{U(\mathfrak{b})} k_\nu$ with $S(\mathfrak{n})$ via the symmetrization map $\mathfrak{g}$ acts by differential operators on $S(\mathfrak{n})$ viewed as a polynomial ring $k(x_{-\alpha})$, which are first order in the $x_{-\alpha}$. Since $M(\nu) \cong U(\mathfrak{n})$ the resulting expressions for $x \in \mathfrak{n}$ are independent of ν and homogeneous in $x_{-\alpha}$.

Taking $y_\alpha = \partial/\partial x_{-\alpha}$ and identifying $M(\nu)^*$ with $k[[y_\alpha]]$ we conclude that $\mathfrak{g}$ acts by first-order differential operators (with coefficients in $k[y_\alpha]$) on $M(\nu)^*$. Moreover any weight vector is necessarily a finite sum (because the y_α have all positive weights) and so $\delta M(\nu)$ is just $k[y_\alpha]$. In addition each $x \in \sigma(\mathfrak{n})$ acts on $\delta M(\nu)$ by derivations. From 2.6 we conclude that

$$\mathrm{ch}\ H^0(\sigma(\mathfrak{m}),\, M(\nu)) = e^\nu \prod_{\alpha \in R^+} (1 - e^{-\alpha})^{-m_\alpha}$$

where $m_\alpha = \dim(\mathfrak{n}/\mathfrak{m})_\alpha$. Applying δ to the resolution in 8.5 we obtain the

COROLLARY.   *Let $L(\mu)$ be a simple integrable highest weight module for a symmetrizable Kac-Moody Lie algebra. Then for any $\mathfrak{h}$ stable subalgebra $\mathfrak{m}$ of $\mathfrak{n}$ one has*

$$\sum_{i=1}^{\infty} (-1)^i\ \mathrm{ch}\ H_i(\mathfrak{m}, L(\mu)) = \frac{\sum_{w \in W} (-1)^{\ell(w)} e^{w \cdot \mu}}{\prod_{\alpha \in R^+} (1 - e^{-\alpha})^{m_\alpha}}$$

with convergence in the Krull topology.

REMARK.   Fix $\nu \in \mathbb{N}R^+$. We conclude that for all μ sufficiently large that $\dim L(\mu)^{\sigma(\mathfrak{m})}_{\mu - \nu}$ equals the coefficient of $e^{-\nu}$ in $\prod(1 - e^{-\alpha})^{-m_\alpha}$. This recovers a result of J. Horvath [26] obtained for $\mathfrak{g}$ semisimple.

8.7.   In general we cannot expect to recover $H_i(\mathfrak{m}, L(\mu))$ from 8.6. Take however the special case when $\mathfrak{m}$ is the nilradical $\mathfrak{m}_{\pi'}$ of a parabolic subalgebra $\mathfrak{p}_{\pi'} \supset \mathfrak{h} + \mathfrak{n}$ with reductive part $\mathfrak{r}_{\pi'}$ defined by a subset π' of the set π of simple roots. That is $\mathfrak{m}$ is the span of the root spaces $kx_{-\alpha} : \alpha \in R'^+ := \mathbb{Z}\pi' \cap R^+$. Let $W_{\pi'}$ denote the subgroup of the Weyl group W generated by the simple reflections $s_\alpha : \alpha \in \pi'$. Set $D_{\pi'} := \{ w \in W \mid w\alpha \in R^+,\ \forall\ \alpha \in \pi' \}$. By ([44, 3.10]) if $w\alpha \in R^-, \alpha \in \pi'$, then $\ell(ws_\alpha) < \ell(w)$. We conclude by induction on reduced length that every $w \in W$ can be expressed uniquely in the form $w = xy : x \in D_{\pi'}, y \in W_{\pi'}$.

Now assume μ integral and dominant. Take $x \in D_{\pi'}$. It is immediate that $x^{-1} \cdot \mu$ is integral and dominant with respect to the (symmetrizable) Kac-Moody Lie algebra $\mathfrak{r}_{\pi'}$ and thus defines a simple integral highest weight module $L_{\pi'}(x^{-1} \cdot \mu)$. One may compute ch $L_{\pi'}(x^{-1} \cdot \mu)$ from 8.5 (replacing R by R'). One obtains

$$\text{ch } L_{\pi'}(x^{-1} \cdot \mu) = \sum_{y \in W_{\pi'}} \frac{(-1)^{\ell(y^{-1})} e^{y^{-1}x^{-1} \cdot \mu}}{\prod_{\alpha \in R'^+} (1 - e^{-\alpha})^{\text{mlt}(\alpha)}},$$

where $\text{mlt}(\alpha)$ denotes the dimension of the root space defined by α.

Comparison with 8.6 shows that one should expect an isomorphism

$$(*) \qquad H_i(\mathfrak{m}_{\pi'}, L(\mu)) = \bigoplus_{x \in D_{\pi'} | \ell(x) = i} L_{\pi'}(x^{-1} \cdot \mu)$$

of $\mathfrak{r}_{\pi'}$ modules. To establish $(*)$ it is enough to show that we have an embedding

$$(**) \qquad H_0(\mathfrak{n}/\mathfrak{m}_{\pi'}, H_i(\mathfrak{m}_{\pi'}, L(\mu)) \hookrightarrow \bigoplus_{x \in D_{\pi'} | \ell(x) = i} k_{x^{-1} \cdot \mu}.$$

of $\mathfrak{h}$ modules.

Define an $\mathfrak{h}$ module map $\varphi_i : \Lambda^i \mathfrak{m} \otimes L \to \Lambda^i \mathfrak{n} \otimes L$ through the embedding $\mathfrak{m} \hookrightarrow \mathfrak{n}$. We have a commutative diagram

$$
\begin{array}{ccccccc}
 & & \uparrow & & \uparrow & & \\
0 & \longrightarrow & \Lambda^i \mathfrak{m} \otimes L & \xrightarrow{\varphi_i} & \Lambda^i \mathfrak{n} \otimes L & \longrightarrow & \\
 & & \uparrow d'_{i+1} & & \uparrow d_{i+1} & & \\
0 & \longrightarrow & \Lambda^{i+1} \mathfrak{m} \otimes L & \xrightarrow{\varphi_{i+1}} & \Lambda^{i+1} \mathfrak{n} \otimes L & \longrightarrow & \\
 & & \uparrow & & \uparrow & & \\
\end{array}
$$

as with exact rows. From this it follows that φ_i factors to $\mathfrak{h}$ module map $\tilde{\varphi}_i : H_i(\mathfrak{m}, L) \to H_i(\mathfrak{n}, L)$. Moreover $\ker \tilde{\varphi}_i = \{a \in \ker d'_i \mid \varphi_i(a) \in \text{Im } d_{i+1}\}/\text{Im } d'_{i+1}$. Given $b = \sum x_j \otimes b_j \in \mathfrak{n}/\mathfrak{m} \otimes (\Lambda^i \mathfrak{m} \otimes L)$ one easily checks that

$$d_{i+1}(b) = \sum x_j b_j + \sum x_j \otimes d_i b_j.$$

Thus if $d_i b_j = 0$, $\forall\, j$ one obtains $\sum x_j b_j \in \mathrm{Im}\ d_{i+1}$. We conclude that $\tilde{\varphi}_i$ factors to a map

$$\bar{\varphi}_i : H_0(\mathfrak{n}/\mathfrak{m},\ H_i(\mathfrak{m}, L)) \longrightarrow H_i(\mathfrak{n}, L).$$

Unfortunately it is not obvious that $\bar{\varphi}_i$ is an embedding unless $i = 0$. Had this been true comparison with 8.6 (taking $\mathfrak{m} = \mathfrak{n}$ there) would have given $(*)$. It does hold however if dim $\mathfrak{n}/\mathfrak{m} = 1$ equivalently if $\pi' = \{\alpha\}$.

8.8 We shall establish $(**)$ above for $\mathfrak{r}_{\pi'}$ any affine Kac-Moody algebra. The key fact is the following

LEMMA. *Let $\mathfrak{g}$ be an affine Kac-Moody Lie algebra. Take $\lambda \in \mathfrak{h}^*$ dominant. Then Verma module $M(\lambda)$ is projective in $\mathcal{O}$. Moreover a surjective map $M \to L \to 0$ in $\mathcal{O}$ with L integrable gives a surjection $H^0(\mathfrak{n}^+, M) \to H^0(\mathfrak{n}^+, L) \to 0$.*

Suppose we have a surjection $M \to M(\lambda) \to 0$ in $\mathcal{O}$. Choose an inverse image $v_\lambda \in M$ of the canonical generator e_λ of $M(\lambda)$. It is enough to show (since $M(\lambda)$ is a free $\mathfrak{n}$ module) that $\mathfrak{n}^+ v_\lambda = 0$. Otherwise M admits a vector $x_{\alpha_i} v_\lambda$ of weight $\lambda + \alpha_i$ and hence a highest weight vector of weight $\nu \neq \lambda$ with $\nu - \lambda \in \mathbb{N}\pi$. Set $\beta = \nu - \lambda$. By primary decomposition of M with respect to the Casimir invariant we can assume $(\nu + \delta, \nu + \delta) = (\lambda + \delta, \lambda + \delta)$ and hence $2(\beta, \lambda + \delta) + (\beta, \beta) = 0$. Yet $\lambda + \delta$ is dominant regular by hypothesis so $(\beta, \lambda + \delta) > 0$. Thus $(\beta, \beta) < 0$ which contradicts that $\mathfrak{g}$ is affine. This proves the first part.

For the second part we recall that by ([44, 9.5, 10.3]) that one has complete reducibility for integrable modules in $\mathrm{Ob}\,\mathcal{O}$. Thus we can assume L simple. Then $L = L(\mu)$ for some μ integral dominant. Then $M(\mu)$ is the projective cover of L and the assertion follows from the first part.

8.9. Retain the hypotheses of 8.7 - 8.8 setting $\mathfrak{m} = \mathfrak{m}_{\pi'}$, $\mathfrak{r} = \mathfrak{r}_{\pi'}$. Let $L(\mu)$ be an integrable highest weight module and set $M_i = \oplus_{\ell(w)=i} M(w \cdot \mu)$. Break the exact sequence

$$0 \longrightarrow L(\mu) \longrightarrow \delta M(\mu) \longrightarrow \cdots \longrightarrow \delta M_i \longrightarrow$$

into short exact sequences

$$0 \longrightarrow X_i \longrightarrow \delta M_i \longrightarrow X_{i+1} \longrightarrow 0$$

where $X_0 = L(\mu)$. Setting $\mathfrak{m}^+ = \sigma(\mathfrak{m})$ we obtain

$$0 \longrightarrow H^0(\mathfrak{m}^+, X_i) \longrightarrow H^0(\mathfrak{m}^+, \delta M_i) \longrightarrow H^0(\mathfrak{m}^+, X_{i+1})$$
$$\longrightarrow H^1(\mathfrak{m}^+, X_i) \longrightarrow 0$$

and

$$H^j(\mathfrak{m}^+, X_{i+1}) \xrightarrow{\sim} H^{j+1}(\mathfrak{m}^+, X_i), \quad \forall\, j \geq 1.$$

From the above we obtain $\mathfrak{r}$ module maps

$$H^0(\mathfrak{m}^+, X_i) \twoheadrightarrow H^i(\mathfrak{m}^+, L(\mu)), \quad H^0(\mathfrak{m}^+, X_i) \hookrightarrow H^0(\mathfrak{m}^+ \delta M_i).$$

From the second map we get an injection

$$H^0(\mathfrak{n}^+/\mathfrak{m}^+, H^0(\mathfrak{m}^+, X_i)) \hookrightarrow H^0(\mathfrak{n}^+/\mathfrak{m}^+, H^0(\mathfrak{m}^+, \delta M_i))$$
$$\hookrightarrow H^0(\mathfrak{n}^+, \delta M_i) = \oplus_{\ell(w)=i} \, k_{w.\mu}$$

of $\mathfrak{h}$ modules. Since $X_i \in \mathrm{Ob}\,\mathcal{O}$ it follows that $\dim U(\mathfrak{n}^+/\mathfrak{m}^+)\xi < \infty$, $\forall\, \xi \in X_i^{\mathfrak{m}^+}$. This shows in particular that $H^0(\mathfrak{m}^+, X_i)$ belongs to the $\mathcal{O}$ category relative to $\mathfrak{r}$. Now assume that $\mathfrak{r}$ is affine. Then by 8.8, the surjection above gives a surjective map

$$H^0(\mathfrak{n}^+/\mathfrak{m}^+, H^0(\mathfrak{m}^+, X_i)) \twoheadrightarrow H^0(\mathfrak{n}^+/\mathfrak{m}^+, H^i(\mathfrak{m}^+, L(\mu))).$$

Taking account of 8.6(*) and the above this implies that every weight of $H_0(\mathfrak{n}/\mathfrak{m}, H_i(\mathfrak{m}, L(\mu)))$ lies in $\{w.\mu \,|\, \ell(w) = i\}$ and is dominant, hence lies in the set $\{w.\mu \,\big|\, w^{-1} \in D_{\pi'}, \ell(w) = i\}$. This establishes (**) of 8.7. We have proved the

THEOREM.   *Let $\mathfrak{g}$ be a symmetrizable Kac-Moody Lie algebra, $\mathfrak{p}_{\pi'} \supset \mathfrak{n} \oplus \mathfrak{h}$ a parabolic subalgebra whose Levi factor $\mathfrak{r}_{\pi'}$ is affine. Let $\mathfrak{m}_{\pi'}$ be the nilradical of $\mathfrak{p}_{\pi'}$. Let $L(\mu)$ be a simple integrable highest weight module. Then for all $i \in \mathbb{N}$ one has an isomorphism*

$$H_i(\mathfrak{m}_{\pi'}, L(\mu)) \cong \bigoplus_{x \in D_{\pi'} \big| \ell(x)=i} L_{\pi'}(x' \cdot \mu)$$

of $\mathfrak{r}_{\pi'}$ modules.

REMARK.   In the special case when $\mathfrak{g}$ is semisimple this recovers a result of B. Kostant [46]. The present proof is very much easier and of course proves a more general result.

REFERENCES

1. J. Alev, T.J. Hodges, and J.-D. Velez, *Fixed rings of the Weyl algebra, $A_1(\mathbb{C})$*, Preprint (1989).
2. H.H. Andersen, *Schubert varieties and Demazure's character formula*, Invent. Math. **79** (1985), 611–618.
3. A. Beilinson and J. Bernstein, *Localization de $\mathfrak{g}$ modules*, Comptes Rendus, Serie I **292** (1981), 15–18.
4. E. Benlulu, *Thèse*, Haifa (1990).
5. I.N. Bernstein, *The analytic continuation of generalized functions with respect to a parameter*, Funct. Anal. Appl. **6** (1972), 26–40.
6. I.N. Bernstein and S.I. Gelfand, *Tensor products of finite and infinite dimensional representations of semisimple, Lie algebras*, Compos. Math. **41** (1980), 245–285.
7. J.-E. Björk and E.K. Ekström. to appear in "Proc. of Colloquium in honour of J. Dixmier", Birkhauser, Boston.
8. A. Borel et al., "Algebraic D-modules," Perspectives in Mathematics, Vol. 2, Academic Press, Boston, 1987.
9. W. Borho, J.-L. Brylinski and R. MacPherson, "Equivariant K theory approach to nilpotent orbits," Progress in Math., Vol. 78, Birkhauser, Boston, 1989.
10. W. Borho, P. Gabriel and R. Rentschler, "Primideal in Einhüllenden auflösbarer Lie-Algebren," Lecture Notes in Math., Vol. 357, Springer-Verlag, Berlin, 1973.
11. W. Borho and H. Kraft, *Über die Gelfand-Kirillov Dimension*, Math. Ann. **220** (1976), 1–24.
12. M. Brion. to appear in "Proc. of Colloquium in honour of J. Dixmier", Birkhauser, Boston.
13. N. Conze, *Algèbres d'opérateurs differentials et quotients des algèbres enveloppantes*, Bull. Math. Soc. France **102** (1974), 379–415.
14. J. Dixmier, "Algèbres enveloppantes," Cahiers Scientifiques XXXVII, Gauthier-Villars, Paris, 1974.
15. S. Fernando, *Lie algebra modules with finite dimensional weight spaces I*, Preprint (1989), Riverside.
16. O. Gabber, *The integrability of the characteristic variety*, Amer. J. Math. **103** (1981), 445–468.
17. O. Gabber, *Equidimensionalité de la variété caracteristique*, Notes rédigé par T. Levasseur, Preprint (1983), Paris 6.
18. O. Gabber and A. Joseph, *On the Bernstein-Gelfand-Gelfand resolution and the Duflo sum formula*, Compos. Math. **43** (1981), 107–131.
19. O. Gabber and V.G. Kac, *On the defining relations of certain infinite dimensional Lie algebras*, Bull. Amer. Math. Soc. **5** (1981), 185–189.
20. V. Ginsburg, *$\mathcal{G}$ modules, Springer's representations and bivariant Chern classes*, Adv. in Math. **61** (1986), 1–48.
21. V. Ginsburg, *Characteristic varieties and vanishing cycles*, Invent. Math. **84** (1986), 327–402.
22. A.B. Goncharov, *Generalized conformal structures on manifolds*, Selecta. Math. Sov. **6** (1987), 307–340.
23. V. Guillemin, D. Quillen and S. Sternberg, *The integrability of characteristics*, Comm. Pure and Applied Math. **23** (1970), 39–77.
24. G. Hochschild and J.-P. Serre, *Cohomology of Lie algebras*, Ann. of Math. **57** (1953), 591–605.
25. T.J. Hodges and S.P. Smith, *Sheaves of non-commutative algebras and the Beilinson-Bernstein equivalence of categories*, Proc. Amer. Math. Soc. **93** (1985), 379–388.

26. J. Horvath, *A dimension theorem for vectors fixed by unipotent groups*, Math. Z. (to appear).

27. R. Hotta, *On Joseph's construction of Weyl group representations*, Tôho-ku Math. J. **36** (1984), 49–74.

28. J.-C. Jantzen, "Einhüllende Algebren halbeinfacher Lie algebren," Ergebnisse der Mathematik, Vol. 3, Band 3, Springer-Verlag, Berlin, 1983.

29. A. Joseph, *A generalization of the Gelfand-Kirillov conjecture*, Amer. J. Math. **99** (1977), 1151–1165.

30. A. Joseph, *A preparation theorem for the prime spectrum of a semisimple Lie algebra*, J. Algebra **48** (1977), 241–289.

31. A. Joseph, *Goldie rank in the enveloping algebra of a semisimple Lie algebra, I-III*, J. Algebra **65** (1980), 269–283, 284–306;;, **73** (1981), 295–326.

32. A. Joseph, *On the variety of a highest weight module*, J. Algebra **88** (1984), 238–278.

33. A. Joseph, *On the associated variety of a primitive ideal*, J. Algebra **93** (1985), 509–523.

34. A. Joseph, *On the Demazure character formula I - II*, Ann. Ec. Norm. Sup. **18** (1985), 381–419;; Compos. Math. **58** (1986), 259–278.

35. A. Joseph, *A sum rule for the scalar factors in the Goldie rank polynomials*, J. Algebra **118** (1988), 276–311.

36. A. Joseph, *A surjectivity theorem for rigid highest weight modules*, Invent. Math. **92** (1988), 567–596.

37. A. Joseph, *The primitive spectrum of an enveloping algebra*, Astérisque **173–174** (1989), 13–53.

38. A. Joseph, *Characteristic polynomials for orbital varieties*, Ann. Ec. Norm. Sup. **22** (1989), 569–603.

39. A. Joseph, *Rings of $\mathfrak{b}$ finite endomorphisms of simple highest weight modules are Goldie*, Preprint (1989).

40. A. Joseph, *The surjectivity theorem, characteristic polynomials and induced ideals*, in "The Orbit Method in Representation Theory," eds. M. Duflo, N.V. Pedersen, and M. Vergne, Birkhauser, Boston, 1990, pp. 85–98.

41. A. Joseph, *Annihilators and associated varieties of unitary highest weight modules*, Preprint (1990).

42. A. Joseph and G. Letzter. (to appear).

43. A. Joseph and T.J. Stafford, *Modules of $\mathfrak{k}$ finite vectors over semisimple Lie algebras*, Proc. Lond. Math. Soc. **49** (1984), 361–384.

44. V.G. Kac, "Infinite dimensional Lie algebras," Cambridge Univ. Press, NY, 1985.

45. M. Kashiwara and T. Kawai, *On the characteristic variety of a holonomic system with regular singularities*, Adv. Math. **34** (1979), 163–184.

46. B. Kostant, *Lie algebra cohomology and generalized Schubert cells*, Ann. Math. **77** (1963), 72–144.

47. G. Krause and T.H. Lenagen, *Growth of algebras and Gelfand-Kirillov dimension*, Research Notes in Mathematics **116** (1985), Pitman, London.

48. T. Levasseur, *Sur la dimension de Krull de l'algèbre enveloppante d'une algèbre de Lie semisimple*, ed. M.-P. Malliavin, Lecture Notes in Math., Vol. 924, in "Séminaire Dubreil-Malliavin," Springer-Verlag, Berlin, 1981, pp. 173–183.

49. T. Levasseur, S.P. Smith and J.T. Stafford, *The minimal nilpotent orbit, the Joseph ideal and differential operators*, J. Algebra **116** (1988), 480–501.

50. T. Levasseur and J.T. Stafford, *Rings of differential operators and classical rings of invariants*, Mem. Amer. Math. Soc. **81** (1989).

51. J.C. McConnell, *Amalgams of Weyl algebras and the $\mathcal{A}(V, \delta, \Delta)$ conjecture*, Invent. Math. **92** (1988), 163–171.

52. J.C. McConnell and J.C. Robson, "Noncommutative Noetherian rings," J. Wiley, NY, 1987.

53. W.M. McGovern, *Dixmier algebras and the orbit method*, in "Proc. of Colloquium in honour of J. Dixmier," Birkhauser, Boston (to appear).

54. W.M. McGovern, *Quantization of nilpotent orbits in complex classical groups*, Preprint (1989).

55. O. Mathieu, *Filtrations of B-modules*, Duke Math. J. **59** (1989), 421–442.

56. C. Moeglin, *Modèle de Whittaker et idéaux primitifs complètement premiers dans les algèbres enveloppantes des algèbres de Lie semisimple complexes II*, Math. Scand. **63** (1988), 5–35.

57. P. Polo, *Un critère d'existence d'une filtration de Schubert*, Comptes Rendus, Serie I **307** (1988), 791–794.

58. I. Pranata, *Structure of Dixmier algebras*, Ph.D. Dissertation (1989), MIT.

59. A. Rocha and N.R. Wallach, *Highest weight modules over graded Lie algebras: Resolutions, filtrations and character formulas*, Trans. Amer. Math. Soc. **277** (1983), 133–162.

60. W. Rossmann, *Equivariant multiplicities on complex varieties*, Astérisque **173–174** (1989), 313–330.

61. M. Rosso, *Groupes quantiques, representations linèaires et applications*, Thèse (1989), Paris 7.

62. S.P. Smith, *Overrings of primitive factor rings of $U(\mathfrak{sl}(2,\mathbb{C}))$*, J. Pure and Applied Algebra (to appear).

63. J.T. Stafford, *Non holonomic modules over Weyl algebras and enveloping algebras* **79** (1985), 619–638.

64. J.T. Stafford, *Endomorphisms of right ideals of the Weyl algebra*, Trans. Amer. Math. Soc. **299** (1987), 623–639.

65. T. Tanisaki, *Characteristic varieties of highest weight modules and primitive quotients*, Advanced Studies in Pure Math. **14** (1988), 1–30.

66. T. Tanisaki, *Harish-Chandra isomorphisms for quantum algebras*, Preprint (1989).

67. P. Tauvel, *Sur les représentations des algèbres de Lie nilpotentes*, Comptes Rendus, Serie I **278** (1974), 977–979.

68. M. Vergne, *Polynômes de Joseph et représentation de Springer*, Ann. Ec. Norm. Sup. (to appear).

69. D.A. Vogan, *The orbit method and primitive ideals for semisimple Lie algebras*, in "Lie algebras and related topics," eds. F.W. Lemire, D. Britten and R.V. Moody, Canad Math. Soc., Vol. 5, for AMS, Providence, RI, 1986.

70. D.A. Vogan, *Dixmier algebras, sheets and representation theory*, in "Proc. of Colloquium in honour of J. Dixmier," Birkhauser, Boston (to appear).

71. D.A. Vogan, *Associated varieties and unipotent representations*, Preprint (1990), M.I.T..

72. M. Zahid, *Les endomorphismes $\mathfrak{k}$-finis des modules de Whittaker*, Bull. Math. Soc. France (to appear).

The Donald Frey Professorial Chair,
The Department of Theoretical Mathematics
The Weizmann Institute of Science
Rehovot 76100, Israel
and
Laboratoire de Mathématiques Fondamentales
Équipe de récherche associée au CNRS
4 Place Jussieu,75232 Paris, Cedex 05, France.

Crossed Products: Characters, Cyclic Homology, and Grothendieck Groups

MARTIN LORENZ

Introduction

This article surveys some recent developments in the study of modules over crossed products $R * G$. As indicated in the title, we focus on three particular techniques, namely

(1) characters or rather their ring theoretic analog, Hattori-Stallings ranks,

(2) cyclic homology and Connes' "Chern characters", an extension of Hattori-Stallings ranks, and

(3) Moody's induction theorem for Grothendieck groups $G_0(R*G)$ where G is a polycyclic-by-finite group.

Correspondingly, the article is divided into three sections whose precise contents are described in their opening paragraphs. Here is a rough overview.

The general point of view of these notes is to treat module theory for crossed products $R * G$ as an extension of group representation theory in a ring theoretic context. Therefore, most of the results discussed here are modelled after, and generalize, certain classical facts about group algebras. For example, in 1.4 we describe a decomposition of the trace group $T(R * G) = R * G/[R * G, R * G]$ into a direct sum with one summand for each conjugacy class of G. This echoes the fact that, for group algebras kG, $T(kG)$ is a k-vector space with basis the set of conjugacy classes of G. The decomposition according to conjugacy classes has been further generalized, at least for skew group rings $R_\alpha G$, to all higher cyclic homology groups $HC_n(R_\alpha G)$ by Feigin and Tsygan [16]. A discussion of their result forms the core of Section 2. Section 3 celebrates Moody's induction theorem, with background and applications. The latter concern the computation of Goldie ranks for $R * G$ and an extension of the classical determination of the number of irreducible representations for finite groups, due to Frobenius and Brauer, to general polycyclic-by-finite groups [24].

Although many results discussed in this article are known, especially all of Section 3, some results seem to be new in the form presented here (e.g.,

Lemma 1.4), others are definitely known but not easily accessible in the literature (e.g., 1.5), and still others amplify or generalize results that have appeared elsewhere (e.g., Theorem 1.6 and Proposition 2.4). Needless to say that these notes do not try to be encyclopedic about their subject. The most glaring omission, perhaps, is a treatment of the Frobenius (or p-th power) map on trace groups in characteristic p. For group algebras, this is a very successful and well-explored technique, and a good reference for this material is Chapter 2 of [31]. Instead, we concentrate on newer techniques that show potential for further development. This is especially true of cyclic homology. For the time being, it requires some effort to construct examples where cyclic homology applies conveniently and which could not be treated equally well by more classical and elementary means, such as Frobenius maps and valuations. However, the subject cyclic homology draws from a great variety of resources, and its use in the present context surely has not been pushed to its limit yet.

Notations and Conventions.

Throughout, rings are associative with identity element and modules are understood to be right modules. Our notation concerning the Grothendieck groups

$$G_0(S) = K_0(\text{finitely generated } S\text{-modules}), \text{ and}$$
$$K_0(S) = K_0(\text{finitely generated projective } S\text{-modules})$$

follows [2]. In particular, $[V]$ denotes the element of $G_0(S)$ (or $K_0(S)$) corresponding to the finitely generated (projective) S-module V, and

$$c: K_0(S) \to G_0(S), \qquad [V] \mapsto [V]$$

is the Cartan map. We use the exponential notation

$$r \mapsto r^g \qquad (g \in G)$$

for actions of a group G. Furthermore,

$$T(G) \qquad \text{is the set of conjugacy classes of the group } G, \text{ and}$$
$$[g] \in T(G) \qquad \text{denotes the conjugacy class of the element } g \in G.$$

§1. Hattori-Stallings ranks

1.1. This section is concerned with trace groups and Hattori-Stallings ranks. Both are special cases (degree zero components) of more elaborate concepts, the former of cyclic homology and the latter of Connes' "Chern character", which will form the subject of the next section. Our point of view here is to view crossed products $S = R * G$ as more complicated versions of group algebras, and their Hattori-Stallings rank functions χ_S as suitable ring theoretic substitutes for characters. The analogy is explained in 1.3(ii) and 1.4. As an application, we extend some well-known facts about modular group algebras of finite p-groups to skew group rings. The corresponding result, in 1.6, stems from [**22**] and [**23**], but the particularly effective use of Hattori-Stallings ranks in the proof of 1.6 comes from T. Hodges [**18**].

With the possible exception of 1.4, virtually everything in this section is known. Morita invariance of trace groups, which is discussed in 1.5, is just a special case of known, albeit less elementary facts about cyclic homology. Morita invariance of Hattori-Stallings ranks is surely well-known, but I am not aware of a reference. Since it is a useful fact, I have included the (easy) proof here.

1.2. Trace groups and Hattori-Stallings ranks (Hattori [**17**], Stallings [**38**]). Let P_S denote a finitely generated projective module over the ring S. Then one can write $P \cong eS^n$ for some n and some idempotent matrix $e = e^2 = (e_{ij}) \in M_n(S)$. We put

$$\chi(P) = \sum_{i=1}^{n} e_{ii} + [S,S] \in S/[S,S] = T(S),$$

where $[S,S]$ denotes the additive subgroup of S that is generated by all Lie commutators $[a,b] = ab - ba$ $(a,b \in S)$. For example,

$$\chi(S^n) = n + [S,S] \in T(S),$$

by using $e = 1_{n \times n}$. The group $T(S) = S/[S,S]$ is called the *trace group of* S. One can check that the above definition of $\chi(P)$ is independent of the choices of n and e made above. It is clear, therefore, that χ is additive, that is

$$\chi(P_1 \oplus P_2) = \chi(P_1) + \chi(P_2),$$

and that χ is well-behaved with respect to induction of modules: If $\varphi : S \to R$ is a ring homomorphism and $\mathrm{Ind}_S^R(P) = P \otimes_S R$ denotes the induced module, then

$$\chi(\mathrm{Ind}_S^R(P)) = T(\varphi)(\chi(P)) \in R/[R,R] = T(R),$$

where $T(\varphi) : T(S) \to T(R)$ sends $s + [S, S]$ to $\varphi(s) + [R, R]$. By the first fact, χ gives rise to a homomorphism

$$\chi = \chi_S : K_0(S) \to T(S) = S/[S, S]$$
$$\cup \qquad \cup$$
$$[P] \mapsto \chi(P)$$

which will be called the *Hattori-Stallings rank function of S*. The second remark implies that, for any ring homomorphism $\varphi : S \to R$, the following diagram commutes

$$
\begin{array}{ccc}
K_0(S) & \xrightarrow{\;\chi_S\;} & T(S) \\
{\scriptstyle \mathrm{Ind}_S^R} \downarrow & & \downarrow {\scriptstyle T(\varphi)} \\
K_0(R) & \xrightarrow[\;\chi_R\;]{} & T(R)
\end{array}
$$

The above definition of χ is actually a special case of a somewhat more general concept. Namely, using the isomorphism $P \otimes_S P^* \cong \mathrm{End}_S(P)$ with $P^* = \mathrm{Hom}_S(P, S)$ and with $p \otimes f \in P \otimes_S P^*$ corresponding to the endomorphism $x \mapsto pf(x)$ of P, one can define a "trace" homomorphism

$$\mathrm{Tr} = \mathrm{Tr}_{P/S} : \mathrm{End}_S(P) \cong P \otimes_S P^* \to T(S)$$
$$\cup \qquad\qquad \cup$$
$$p \otimes f \mapsto f\;(p) + [S, S]$$

The image of $\mathrm{id}_P \in \mathrm{End}_S(P)$ under Tr is identical with $\chi(P)$. We will not pursue this aspect much further in these notes (but see 1.3 and 1.5 below). Furthermore, the definition of χ extends in a straghtforward manner to modules having finite resolutions by finitely generated projectives (cf. Bass [3]).

1.3. Classical examples. (i) *Commutative Rings.* If $S = k$ is a commutative ring, then $T(S) = k$ and, for any finitely generated projective k-module P, the above homomorphism

$$\mathrm{Tr}_{P/k} : \mathrm{End}_k(P) \to k$$

associates to each endomorphism of P its trace in the classical sense (e.g., [5, chap. II §4 no. 3]).

(ii) *Group Algebras.* Let $S = kG$ denote the group algebra of the group G over the commutative ring k. Then the trace group $T(S)$ is usually identified with the free k-module on the set $T(G)$ of all conjugacy classes of G. Explicitly, the identification is as follows.

$$T(kG) = kG/[kG, kG] \quad \to k^{(T(G))}$$

$$\cup\!\!\!| \qquad\qquad\qquad \cup\!\!\!|$$

$$\sum_{x\in G} k_x x + [kG, kG] \mapsto (\sum_{x\in[g]} k_x)_{[g]\in T(G)}$$

Viewing $k^{(T(G))}$ as the set of finitely supported functions $T(G) \to k$, one can interpret Hattori-Stallings ranks $\chi(P)$ as functions $G \to k$ which are constant on conjugacy classes of G and vanish on almost all conjugacy classes. This gives meaning to the notation

$$\chi(P)(g) \in k \qquad (g \in G).$$

The following classical fact explains why Hattori-Stallings ranks can be viewed as an appropriate ring theorctical version of the characters from finite group representation theory. Recall that, for a finite group G and a finitely generated projective kG-module P, the *ordinary character of P*, $\mathrm{ch}(P)$, is given by

$$\mathrm{ch}(P) = (\mathrm{Tr}_{P/k}(g_P))_{[g]\in T(G)} \in k^{(T(G))}.$$

Here, g_P stands for the k-endomorphism of P that is given by the action of $g \in G$ and $\mathrm{Tr}_{P/k}$ is as in 1.2 or 1.3(i).

LEMMA (Hattori [17]). *Let G be a finite group and let P be a finitely generated projective kG-module. Then*

$$\mathrm{ch}(P)(g) = |\mathbb{C}_G(g)| \cdot \chi(P)(g^{-1}) \qquad (g \in G),$$

where $\mathbb{C}_G(g)$ is the centralizer of g in G.

The "renormalization factor" $|\mathbb{C}_G(g)|$ here reflects the fact that Hattori-Stallings ranks behave perfectly well with respect to induction (1.1), other than ordinary characters where group indices complicate matters somewhat.

1.4. The trace group of a crossed product. Our aim here is to give a description of the trace group $T(S)$ of a crossed product $S = R * G$ which generalizes the one for group algebras kG in 1.3(ii). We will decompose $T(R * G)$ into a direct sum

$$T(R * G) = \bigoplus_{[g] \in T(G)} T(R * G)_{[g]}$$

whose components $T(R*G)_{[g]}$ in general have a somewhat more complicated form than in the case of group algebras where $T(kG)_{[g]} = k$ for all $[g]$. The following **notation** will be kept throughout.

> $S = R * G$ will denote a crossed product of the group G over the ring R. So $S = \bigoplus_{g \in G} S_g$ is a G-graded ring, with $S_1 = R$, so that each homogeneous component S_g contains an invertible element of S. We fix such an element $\bar{g} \in S_g$ for each $g \in G$. Moreover, for each $g \in G$, we put
>
> $G_g = \mathbb{C}_G(g)/\langle g \rangle,$
>
> $S_{[g]} = \bigoplus_{x \in [g]} S_x \subseteq S,$
>
> $V_g = S_g/[S_g, R] = H_0(R, S_g),$ and
>
> $V_{[g]} = S_{[g]}/[S_{[g]}, R] = H_0(R, S_{[g]}).$

Here $H_0(R, .)$ denotes the 0-th Hochschild homology group of R with coefficients in the specified (R, R)-bimodule (cf. [**26**], p. 289). Recall also that, for any $\mathbb{Z}G$-module V, $H_0(G, V) = V/\sum_{x \in G} V(1 - x)$ is the group of G-coinvariants in V.

LEMMA.   (a) *For each $g \in G$, G_g acts on V_g via*

$$(s + [S_g, R])^x = \bar{x}^{-1} s \bar{x} + [S_g, R] \qquad (s \in S_g, x \in \mathbb{C}_G(g)).$$

Similarly, G acts on $V_{[g]}$, and $V_{[g]} \cong \mathrm{Ind}_{\mathbb{C}_G(g)}^G (V_g)$ as $\mathbb{Z}G$-modules.

(b) $T(R * G) \cong \bigoplus_{[g] \in T(G)} H_0(G, V_{[g]}) \cong \bigoplus_{[g] \in T(G)} H_0(G_g, V_g).$

PROOF: (a) First, for each $x \in \mathbb{C}_G(g)$, one has $\bar{x}^{-1} S_g \bar{x} = S_g$ and $\bar{x}^{-1}[S_g, R]\bar{x} = [S_g, R]$. Hence conjugation with $\bar{x}$ yields an automorphism $\bar{x}^{-1}(.)\bar{x}$ of V_g. Moreover, if $x, y \in \mathbb{C}_G(g)$ then $\bar{x}\bar{y} = \overline{xy}u$ for some unit $u \in R$. One computes

$$\bar{y}^{-1}(\bar{x}^{-1} s \bar{x})\bar{y} + [S_g, R] = [u^{-1}, \overline{xy}^{-1} s \overline{xy} u] + \overline{xy}^{-1} s \overline{xy} + [S_g, R]$$
$$= \overline{xy}^{-1} s \overline{xy} + [S_g, R].$$

Therefore, the map $x \mapsto \overline{x}^{-1}(.)\overline{x}$ defines an action of $\mathbb{C}_G(g)$ on V_g. Finally, since $\overline{g} \in S_g$, one has for each $s \in S_g$,

$$\overline{g}^{-1}s\overline{g} - s = [\overline{g}^{-1}s, \overline{g}] \in [S_g, R].$$

Thus $\langle g \rangle$ acts trivially on V_g and so $G_g = \mathbb{C}_G(g)/\langle g \rangle$ acts. Similarly, G acts on $V_{[g]}$. Furthermore,

$$V_{[g]} = (\underset{x \in [g]}{\oplus} S_x)/[\underset{x \in [g]}{\oplus} S_x, R] \cong \underset{x \in [g]}{\oplus} S_x/[S_x, R]$$
$$= \underset{x \in [g]}{\oplus} V_x$$

and

$$(V_x)^y = V_{x^y} \qquad (x \in [g], \ y \in G).$$

Thus G permutes the components V_x of $V_{[g]}$ transitively, whence $V_{[g]} \cong \mathrm{Ind}^G_{\mathbb{C}_G(g)}(V_g)$.

(b) By part (a) and Shapiro's Lemma, $H_0(G, V_{[g]}) \cong H_0(\mathbb{C}_G(y), V_g) = H_0(G_g, V_g)$. Thus we can concentrate on the first isomorphism. For each $x, y \in G$ and $a, b \in R$, one computes

$$[a\overline{x}\,\overline{y}^{-1}, \overline{y}b] = a\overline{x}b - \overline{y}(ba\overline{x})\overline{y}^{-1} = [a\overline{x}, b] + (ba\overline{x})^{1-\overline{y}^{-1}}.$$

Therefore, $[S_{xy^{-1}}, S_y] \subseteq [S_x, R] + S_x^{1-\overline{y}^{-1}}$ and so

$$[S, S] = \sum_{x,y \in G} [S_{xy^{-1}}, S_y] \subseteq \sum_{x,y \in G} ([S_x, R] + S_x^{1-\overline{y}^{-1}})$$
$$= \underset{[g] \in T(G)}{\oplus} ([S_{[g]}, R] + \sum_{y \in G} S_{[g]}^{1-\overline{y}^{-1}}).$$

Since each term in the last expression is clearly contained in $[S, S]$, equality holds here. Therefore,

$$T(S) = S/[S, S] \cong \underset{[g] \in T(G)}{\oplus} (S_{[g]}/([S_{[g]}, R] + \sum_{y \in G} S_{[g]}^{1-\overline{y}^{-1}}))$$
$$= \underset{[g] \in T(G)}{\oplus} H_0(G, V_{[g]}),$$

which completes the proof. $\square$

An *explicit description of the isomorphism in (b)* is as follows. For each $g \in G$, fix a transversal T_g for $\mathbb{C}_G(g)$ in G. So $G = \underset{t \in T_g}{\dot{\cup}} \mathbb{C}_G(g)t$. Then,

writing elements $s \in S = R * G = \underset{x \in G}{\oplus} S_x$ in the form $s = \sum_{x \in G} s_x =$
$\sum_{[g] \in T(G)} \sum_{t \in \mathcal{T}_g} s_{g^t}$ $(s_x \in S_x)$, the isomorphism is given by

$$T(S) = S/[S,S] \overset{\cong}{\longrightarrow} \underset{[g] \in T(G)}{\oplus} H_0(G_g, V_g)$$

$$\cup \qquad\qquad\qquad \cup$$

$$\sum_{[g] \in T(G)} \sum_{t \in \mathcal{T}_g} s_{g^t} + [S,S] \mapsto (\overline{\sum_{t \in \mathcal{T}_g} t s_{g^t} t^{-1}})_{[g] \in T(G)},$$

where $\overline{}$ on the right denotes images in $H_0(G_g, V_g)$ of elements of S_g.

In the special case where $S = RG$ is the *group ring* of G over R the actions on the modules V_g and $V_{[g]}$ are trivial, and $V_g = R_g/[R_g, R] \cong R/[R,R] = T(R)$. Thus

$$T(RG) \overset{\cong}{\longrightarrow} T(R)^{(T(G))}$$

$$\cup \qquad\qquad\qquad \cup$$

$$\sum_{x \in G} r_x x + [RG, RG] \mapsto (\sum_{x \in G} r_x + [R,R])_{[g] \in T(G)}.$$

For commutative R, this is the classical isomorphism described in 1.3(ii).

1.5. Morita invariance. Goal of this subsection is to show that $K_0(R)$, the trace group $T(R)$, and the Hattori-Stallings rank function χ_R are Morita invariants of a given ring R. We remark that, more generally, the full cyclic homology $HC_*(R)$ of R is known to be Morita invariant (e.g., Seibt [**37**]), but we will concentrate on the completely elementary case $T(R) = HC_0(R)$.

Recall that two rings R and S are *Morita equivalent* if and only if there exists a finitely generated projective generator P_R for R with $S \cong \operatorname{End}(P_R)$. In this case, P has the structure of an (S, R)-bimodule and the functor $(.) \otimes_S P_R$ yields an equivalence between the categories of right S-modules and right R-modules. Since all constructions respect isomorphisms, we may assume that

$$S = \operatorname{End}(P_R),$$

and we will do so in the following. We will also put

$$P^* = \operatorname{Hom}_R(P, R).$$

Then P^* is an (R, S)-bimodule, and one has the following bimodule isomorphisms.

$$_SP \underset{R}{\otimes} P_S^* \cong S = \text{End}(P_R)$$

$$\uplus \qquad\qquad \uplus$$

$$p \otimes f \mapsto (x \mapsto pf(x))$$

and

$$_RP^* \underset{S}{\otimes} P_R \cong R$$

$$\uplus \qquad \uplus$$

$$f \otimes p \mapsto f(p).$$

Writing the image of $p \otimes f$ in S as pf and the image of $f \otimes p$ in R as fp,
one observes the following associativity conditions, for all $p, p' \in P$ and
$f, f' \in P^*$,

$$(pf)p' = p(fp') \qquad \text{and} \qquad (fp)f' = f(pf').$$

An isomorphism $\varphi = \varphi_P : K_0(S) \to K_0(R)$ is provided by $(.) \underset{S}{\otimes} P_R$. As to
trace groups, it is a simple matter to check that the map

$$\psi = \psi_P : T(S) = S/[S, S] \to T(R) = R/[R, R]$$

$$\uplus \qquad\qquad\qquad \uplus$$

$$\sum p_i f_i + [S, S] \mapsto \sum f_i p_i + [R, R]$$

is an isomorphism. A somewhat simpler description of ψ is as follows. *Fix
$p_i \in P$ and $f_i \in P^*$ so that $1 = \sum p_i f_i$ holds in S.* (This amounts to
choosing dual bases for P.) Then, for all $s \in S$, one has $s = \sum (sp_i)f_i = \sum p_i(f_i s)$, and hence

$$\psi_P(s + [S, S]) = \sum f_i(sp_i) + [R, R] = \sum (f_i s)p_i + [R, R].$$

Standard Example. R and $S = M_n(R)$ are Morita equivalent, with $_SP_R = R_{n \times 1}$, the set of columns of length n with entries from R. Here, $_RP_S^* = R_{1 \times n}$, the rows of length n with entries from R, and the isomorphisms
$P \underset{R}{\otimes} P^* \cong S$ and $P^* \underset{S}{\otimes} P \cong R$ are given by matrix multiplication. The
elements p_i and f_i can be taken to be the canonical basis elements of P
and P^* (1 in the i-th position and 0's elsewhere). One obtains the formula

$$\psi((r_{ij}) + [S, S]) = \sum_{i=1}^{n} r_{ii} + [R, R] = \text{Trace}(r_{ij}) + [R, R].$$

Finally, returning to the general setting, we claim that Hattori-Stallings ranks respect the isomorphisms φ_P and ψ_P, that is the following diagram commutes.

$$
\begin{array}{ccc}
K_0(S) & \xrightarrow{\;\chi_S\;} & T(S) \\
{\scriptstyle\cong}\Big\downarrow{\scriptstyle\varphi_P} & & {\scriptstyle\cong}\Big\downarrow{\scriptstyle\psi_P} \\
K_0(R) & \xrightarrow[\;\chi_R\;]{} & T(R)
\end{array}
$$

PROOF: Let X_S be finitely generated projective and put $F(X)_R = X \otimes_S P_R$. So $\varphi_P([X]) = [F(X)] \in K_0(R)$. Write $\mathrm{id}_X = \sum x_j \otimes g_j \in X \otimes_S X^* \cong \mathrm{End}(X_S)$. Then

$$\chi_S([X]) = \sum g_j(x_j) + [S, S] \in T(S).$$

As above, fix $p_i \in P$ and $f_i \in P^*$ with $1 = \sum p_i f_i \in S$. Then

$$(\psi_P \circ \chi_S)([X]) = \sum_{i,j} f_i(g_j(x_j)p_i) + [R, R].$$

To compute $(\chi_R \circ \varphi_P)([X]) = \chi_R([F(X)])$, note that $F(X)^* \cong P^* \otimes_S X^*$, with $f \otimes g \in P^* \otimes_S X^*$ corresponding to the map $x \otimes p \mapsto f(g(x)p)$ ($x \in X, p \in P$). Therefore, $\mathrm{End}(F(X)_R) \cong F(X) \otimes_R F(X)^* \cong (X \otimes_S P) \otimes_R (P^* \otimes_S X^*)$, with $(x \otimes p) \otimes (f \otimes g)$ corresponding to the endomorphism $x' \otimes p' \mapsto (x \otimes p) \cdot f(g(x')p)$ of $F(X) = X \otimes_S P$. Using our above choices for p_i, f_i, x_j, g_j one readily checks that $\mathrm{id}_{F(X)} \in \mathrm{End}(F(X)_R)$ corresponds to $\sum_{i,j} (x_j \otimes p_i) \otimes (f_i \otimes g_j) \in (X \otimes_S P) \otimes_R (P^* \otimes_S X^*)$. Therefore,

$$\chi_R([F(X)]) = \sum_{i,j} f_i(g_j(x_j)p_i) + [R, R]$$
$$(\psi_P \circ \chi_S)([X]).$$

This proves commutativity of the diagram. $\square$

1.6. An application to skew group rings. In this paragraph, we use Hattori-Stallings ranks to study the structure of the Grothendieck groups $G_0(S)$ and $K_0(S)$ for certain skew group rings $S = R_\alpha G$. Here, $\alpha : G \to$

$\mathrm{Aut}(R)$ is a group homomorphism and the multiplication in S is given by the rule

$$(r_1g_1) \cdot (r_2g_2) = r_1r_2^{\alpha(g_1^{-1})}g_1g_2 \qquad (r_i \in R, g_i \in G).$$

Additively $R_\alpha G$ looks like the ordinary group ring RG. The ring R has a right S-module structure which extends the right regular R-module structure and is defined by

$$r \cdot \sum_{g \in G} r_g g = \sum_{g \in G} (rr_g)^{\alpha(g)}.$$

We will denote this module by R_S.

The operation of G on R via α induces G-actions on $K_0(R)$ and $G_0(R)$ which can be thought of as the induction maps Ind_R^R for G_0 and K_0 that are given by the automorphisms $\alpha(g) \in \mathrm{Aut}(R)$. Therefore, it makes sense to consider the G-coinvariants $H_0(G, K_0(R))$ and $H_0(G, G_0(R))$. Since we are dealing with G_0, we will assume that the rings considered are right Noetherian. Recall that $c : K_0(S) \to G_0(S)$ denotes the Cartan map. The result below generalizes classical (and easy) facts about modular group algebras of finite p-groups.

THEOREM ([22], [23]). *Let* $S = R_\alpha G$ *be a skew group ring with* R *right Noetherian. Assume that*

(1) G *is a finite p-group* $\neq \langle 1 \rangle$, *for some prime p, and* $p \cdot R = \{0\}$.

Then $H_0(G, K_0(R)) \underset{\mathbb{Z}}{\otimes} \mathbb{Z}[1/|G|] \cong K_0(S) \underset{\mathbb{Z}}{\otimes} \mathbb{Z}[1/|G|]$, *and similarly for G_0 in place of K_0. Assume that, in addition to (1), the following conditions are satisfied.*

(2) $K_0(R) = \langle [R] \rangle$, *and*

(3) $1 \notin [S, S]$ *(or, equivalently,* $1 \notin [R, R] + \sum_{g \in G} R^{1 - \alpha(g)}$*).*

Then $c(K_0(S)) \subseteq p \cdot \langle [R_S] \rangle + \mathrm{ann}_{G_0(S)}(|G|)$ *holds in* $G_0(S)$, *where* $\mathrm{ann}_{G_0(S)}(|G|) = \{\alpha \in G_0(S) \mid |G| \cdot \alpha = 0\}$. *In case $K_0(S)$ is torsion-free one has*

$$K_0(S) = \langle [S] \rangle \qquad and \qquad c(K_0(S)) = |G| \cdot \langle [R_S] \rangle.$$

SKETCH OF PROOF: We first recall a few general facts about skew group rings $S = R_\alpha G$ of finite groups G. Details can be found in [23]. Assume that R is an algebra over a field k and G acts on R by k-algebra automorphisms. Then $K_0(S)$ and $G_0(S)$ are modules over the ring $G_0(kG)$, where the multiplication in $G_0(kG)$ as well as the operations of $G_0(kG)$ on $K_0(S)$ and $G_0(S)$ are afforded by the tensor product $\underset{k}{\otimes}$. The Cartan map $c : K_0(S) \to G_0(S)$ is a module homomorphism, and the map

$$\mathrm{Ind}_R^S \circ \mathrm{Res}_R^S : G_0(S) \to G_0(R) \to G_0(S)$$

is identical with the action of $[kG] \in G_0(kG)$ on $G_0(S)$. Furthermore, the map $\mathrm{Ind}_R^S : G_0(R) \to G_0(S)$ factors through the canonical map $G_0(R) \to H_0(G, G_0(R))$. Letting $\overline{\mathrm{Ind}}_R^S : H_0(G, G_0(R)) \to G_0(S)$ denote the map provided by Ind_R^S, and putting $\overline{\mathrm{Res}}_R^S : G_0(S) \xrightarrow{\mathrm{Res}} G_0(R) \xrightarrow{\mathrm{can.}} H_0(G, G_0(R))$ one easily sees that

$$\overline{\mathrm{Res}}_R^S \circ \overline{\mathrm{Ind}}_R^S : H_0(G, G_0(R)) \to G_0(S) \to H_0(G, G_0(R))$$

is just multiplication with $|G|$ on $H_0(G, G_0(R))$. These facts also hold verbatim with G_0 replaced by K_0.

Assuming now that (1) holds we can take $k = \mathbb{F}_p$ in the above, and one also has $G_0(kG) \cong \mathbb{Z}$, with $[kG] \in G_0(kG)$ corresponding to $|G| \in \mathbb{Z}$. Therefore the foregoing implies that

$$\overline{\mathrm{Ind}}_R^S \circ \overline{\mathrm{Res}}_R^S : G_0(S) \to H_0(G, G_0(R)) \to G_0(S)$$

is multiplication with $|G|$ on $G_0(S)$. This proves the first isomorphism in the theorem, with G_0 instead of K_0, but the above argument works for K_0 as well. We also note the following consequence: in $G_0(S)$ we have

$$|G| \cdot [R_S] = \overline{\mathrm{Ind}}_R^S \circ \overline{\mathrm{Res}}_R^S([R_S]) = [R \underset{R}{\otimes} S] = [S].$$

Now assume that (2) also holds. Then G acts trivially on $K_0(R)$ and, for any finitely generated projective S-module P_S, we have $\mathrm{Res}_R^S([P_S]) = [P_R] = n \cdot [R]$ for some $n \in \mathbb{Z}$. Therefore,

(*) $$|G| \cdot [P_S] = \mathrm{Ind}_R^S \circ \mathrm{Res}_R^S([P_S]) = n \cdot [S],$$

and this equation can be interpreted in $K_0(S)$ as well as in $G_0(S)$. In the latter form it can be combined with the first equation to yield

$$|G| \cdot [P_S] = |G| \cdot n \cdot [R_S]$$

in $G_0(S)$. So $[P_S] - n[R_S] \in \mathrm{ann}_{G_0(S)}(|G|)$. To prove the inclusion $c(K_0(S)) \subseteq p \cdot \langle [R_S] \rangle + \mathrm{ann}_{G_0(S)}(|G|)$ it suffices to show that p divides n.

For this, read $(*)$ in $K_0(S)$ and apply the Hattori-Stallings rank function $\chi = \chi_S$ to obtain the following equation in $T(S) = S/[S,S]$:

$$0 = |G| \cdot \chi(P_S) = n \cdot \chi(S) = n \cdot (1 + [S,S]).$$

Here, the first equality follows from the fact that $T(S)$ is a k-vector space. The same fact combined with (3) further implies that $p|n$, as we have claimed.

Finally, assume that, in addition to (1), (2), (3), $K_0(S)$ is torsion-free. (It follows from (1) and (2) that all torsion elements of $K_0(S)$ are annihilated by $|G|$.) Then Res_R^S embeds $K_0(S)$ into $K_0(R) \cong \mathbb{Z}$ and so $K_0(S) \cong \mathbb{Z}$. Say $K_0(S) = \langle \sigma \rangle$ and $[S] = s\sigma$. It suffices to show that $|s| = 1$. By $(*)$, $|G| \cdot \sigma = n[S] = ns\sigma$ for some n, and so $|G| = ns$. Thus if $|s| > 1$ then $|s|$ is a positive p-power, and hence $[S] \in p \cdot K_0(S)$. But $p \cdot K_0(S) \subseteq \mathrm{Ker}\,\chi_S$ whereas $[S] \notin \mathrm{Ker}\,\chi_S$, by (3). Therefore we must have $|s| = 1$, and the proof is complete.   □

COROLLARY. *Let G be a finite p-group acting on a division ring D with char $D = p$ and put $S = D_\alpha G$, where $\alpha : G \to \mathrm{Aut}(D)$ is the given action. Suppose that*

$$1 \notin [D,D] + \sum_{g \in G} D^{1-\alpha(g)}.$$

Then $K_0(S) = \langle [S] \rangle$, $G_0(S) = \langle [D_S] \rangle$, and $c(K_0(S)) = |G| \cdot \langle [D_S] \rangle$.

PROOF: Note that (2) is clearly satisfied for $R = D$, and $K_0(S)$ and $G_0(S)$ are torsion-free because S is Artinian. The assertions about $K_0(S)$ and $c(K_0(S))$ follow. As to $G_0(S)$, recall from the above proof that

$$\mathrm{Ind}_D^S \circ \mathrm{Res}_D^S : G_0(S) \to G_0(D) = \langle [D] \rangle \to G_0(S)$$

is multiplication with $|G|$ on $G_0(S)$ and has image $\langle [S] \rangle = |G| \cdot \langle [D_S] \rangle$. Since $G_0(S)$ is torsion-free, the fact about $G_0(S)$ also follows.   □

In the situation of the corollary, $S = D_\alpha G$ cannot be semisimple (if $G \neq \langle 1 \rangle$), for in this case the Cartan map is an isomorphism. On the other hand, if G acts by outer automorphisms on D then $S = D_\alpha G$ is simple Artinian (cf. [28, Theorem 2.3]). Thus this case is implicitly ruled out by our assumptions.

For further explicit examples of skew rings with (1), (2), and (3), we refer the reader to [22], [32]. Here we just point out the following consequence of the theorem which was the original motivation for proving it.

COROLLARY [22]. *Let $S = R_\alpha G$ be given with (1), (2) and (3). Then S is not Morita equivalent to a Noetherian domain.*

PROOF: Suppose S is Morita equivalent to T and the equivalence is explicitly given by $(.) \underset{S}{\otimes} P_T$ (1.5). Then $(.) \underset{S}{\otimes} P_T$ yields isomorphisms $K_0(S) \cong K_0(T)$ and $G_0(S) \cong G_0(T)$ which make the following diagram commute

$$
\begin{array}{ccc}
K_0(S) & \overset{c}{\to} & G_0(S) \\
\| \wr & & \| \wr \\
K_0(T) & \underset{c}{\to} & G_0(T).
\end{array}
$$

The theorem therefore implies that $c(K_0(T)) \subseteq p \cdot G_0(T) + \mathrm{ann}_{G_0(T)}(|G|)$. Now, if T is a Noetherian domain and $\rho : G_0(T) \to \mathbb{Z}$ is Goldie's reduced rank function (see 3.5) then $\rho(T) = 1$. But $[T] \in c(K_0(T))$ implies that $\rho(T)$ must be divisible by p, contradiction. □

§2. Cyclic homology

2.1. Our goal in this section is to explain how higher cyclic homology groups of an algebra S can be used to derive information about the Hattori-Stallings rank function χ_S of S. The basic strategy employed here has previously been used, amongst others, by Marciniak [27] and Eckmann [11]. The new ingredient here is the (quite complicated) computation of the cyclic homology for skew group rings by Feigin and Tsygan [16]. Their work extends earlier results of Burghelea [8] on group rings.

   After listing some basic facts about cyclic homology in 2.2, we describe the results of Feigin and Tsygan in 2.3 in sufficient detail so as to allow applications to Hattori-Stallings ranks for skew group rings. This is done in 2.4, while the final section 2.5 contains a fairly straightforward, and essentially known, application of this material to the non-existence of idempotents in group rings.

2.2. Cyclic homology and Connes' Chern character. (Loday-Quillen [20], Karoubi [19], Seibt [37]). Let S be an algebra over a commutative ring k. The cyclic homology groups $HC_n(S)$ $(n \geq 0)$ of S are usually defined by taking homology of the total complex $\mathcal{D}(S) = \mathrm{Tot}\,\mathcal{C}(S)$ of a suitable double complex $\mathcal{C}(S) = (C_{p,q})_{p,q \geq 0}$ with $C_{p,q} = S \underset{k}{\otimes} \ldots \underset{k}{\otimes} S = S^{\otimes(q+1)}$ $(q + 1$ factors). For the details of the definition, we refer to the above references. Here we just summarize some basic facts which will provide sufficient background for our purposes.

(i) $HC_0(S) = T(S) = S/[S, S]$ is the trace group of S.

(ii) Let $H_n(S) = H_n(S, S)$ denote the n-th Hochschild homology group of S with coefficients in the (S, S)-bimodule S (MacLane [**26**], p. 288). (When S is flat over k then $H_n(S) = \operatorname{Tor}_n^{S^e}(S, S)$, where $S^e = S \underset{k}{\otimes} S^{op}$ is the enveloping algebra of S.) Cyclic and Hochschild homology are related by a fundamental long exact sequence of the form

$$\cdots \to H_n(S) \to HC_n(S) \overset{\sigma}{\to} HC_{n-2}(S) \to H_{n-1}(S) \to \cdots,$$

the so-called *Connes-Gysin sequence*. The map σ is called the *shift map* or *periodicity map*. We illustrate the use of this sequence in the following

Example: Suppose that S is a separable k-algebra, that is S is projective as module over $S^e = S \underset{k}{\otimes} S^{op}$. Then $H_n(S) = 0$ holds for all $n > 0$. Therefore the Connes-Gysin sequence implies that

$$HC_n(S) = \begin{cases} 0 & \text{for } n \text{ odd} \\ T(S) & \text{for } n \text{ even.} \end{cases}$$

We further remark that S has a decomposition $S \cong [S, S] \oplus T(S)$ as k-modules (see Hattori [**16**]).

(iii) The importance of cyclic homology for our purposes comes from the fact that there is a sequence of homomorphisms

$$\mathrm{Ch}_0^n : K_0(S) \to HC_{2n}(S) \qquad (n \geq 0),$$

the *Chern characters* of A. Connes, such that the following diagram commutes

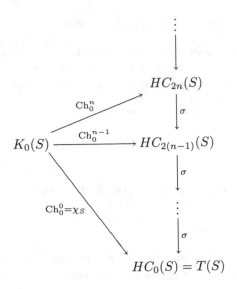

Here, $\text{Ch}_0^0 : K_0(S) \to HC_0(S) = T(S)$ is identical with the Hattori-Stallings rank function χ_S that was discussed in detail in Section 1.

2.3. The cyclic homology of skew group rings (Feigin and Tsygan [**16**]). Let G be a group, let R be an algebra over a commutative ring k, and let $S = R_\alpha G$ be a skew group ring of G over R, where $\alpha : G \to \text{Aut}_{k-\text{alg}}(R)$ is a given homomorphism. Then Feigin and Tsygan have constructed a decomposition

$$HC_n(S) = \bigoplus_{[g]\in T(G)} HC_n(S)_{[g]} \qquad (n \geq 0)$$

which is respected by the shift map, that is $\sigma(HC_n(S)_{[g]}) \subseteq HC_{n-2}(S)_{[g]}$. Moreover, for $n = 0$ (trace group), the decomposition is the one described for crossed products in 1.4 whereas, in general, $HC_n(S)_{[g]}$ is given as the abutment of a certain spectral sequence whose precise form depends upon the element $g \in G$. Instead of discussing this in general, we present some reasonably accessible special cases.

(a) *Conjugacy classes of infinite order.* The E^2-term of the spectral sequence can be computed via grou   homology of the group $G_g = \mathbf{C}_G(g)/\langle g \rangle$

(notation as in 1.4). If the element $g \in G$ has infinite order then

$$E^2_{p,q} \cong H_p(G_g, H_q(R, Rg)),$$

with $H_q(R, Rg)$ denoting the q-th Hochschild homology group of R with coefficients in the (R, R)-bimodule $Rg = \{rg \mid r \in R\} = S_g$, the g-homogeneous component of $S = R_\alpha G$.

Example 1. If the algebra R is separable then each Rg is projective as R^e-module, as R is. Therefore, $H_q(R, Rg) = 0$ holds for all $q > 0$, and it follows that $HC_n(S)_{[g]} \cong E^2_{n,0}$. So

$$HC_n(S)_{[g]} \cong H_n(G_g, V_g) \qquad (n \geq 0),$$

the n-th homology group of G_g with coefficients in $V_g = Rg/[Rg, R] = H_0(R, Rg)$ (notation as in 1.4).

Example 2. Assume that R is k-flat, and

$$\dim R := \text{proj.} \dim_{R^e}(R) = d < \infty, \text{and}$$
$$hd_k(G_g) = h < \infty,$$

where $hd_k(.)$ denotes homological dimension over k. Then $H_q(R, .) = \text{Tor}^{R^e}_q(R, .) = 0$ for all $q > d$ and $H_p(G_g, V) = 0$ for all $p > h$ and all kG_g-modules V. Consequently, $E^2_{p,q} = 0$ if either $p > h$ or $q > d$, whence

$$HC_n(S)_{[g]} = 0 \text{ for } n > h + d.$$

(b) *The identity class.* One has

$$HC_n(S)_{[1]} \cong \mathbb{H}_n(G, \mathcal{D}(R)),$$

the n-th hyperhomology group of G with coefficients in the complex of kG-modules $\mathcal{D}(R) = (\overset{m+1}{\underset{i=1}{\oplus}} R^{\otimes i})_{m \geq 0}$ from 2.2. Here G acts diagonally on each $R^{\otimes i}$, that is $(r_1 \otimes \cdots \otimes r_i)^g = r_1^g \otimes \cdots \otimes r_i^g$.

Hyperhomology is discussed in great generality in Chapter XVII of [9], and more specifically for groups in [7, p. 168ff]. Here we just note that if k is a field (or a finite direct product of fields) and $\mathcal{D} = (D_m)_{m \geq 0}$ is a complex of kG-modules with trivial G-action then the hyperhomology of G in $\mathcal{D}$ has the form

$$\mathbb{H}_n(G, \mathcal{D}) \cong \underset{p+q=n}{\oplus} H_p(\mathcal{D}) \underset{k}{\otimes} H_q(G, k),$$

[7, (5.4) on p. 169].

(c) *The remaining classes of finite order.* Feigin and Tsygan also determine $HC_n(S)_{[g]}$ for arbitrary elements $g \in G$ of finite order, provided $k \supseteq \mathbb{Q}$. The description is given in terms of hyperhomology groups of $G_g = \mathbb{C}_G(g)/\langle g \rangle$ with coefficients in a certain complex depending upon g. In the special case when $S = RG$ is the *group ring*, that is $\alpha(g) = \mathrm{id}_R$ for all $g \in G$, the complex in question is $\mathcal{D}(R)$, as above, with trivial G_g-operation. In view of our above remarks about hyperhomology we deduce the following result.

Assume that k is a field of characteristic 0 (or a finite direct product of such fields). Then for all $g \in G$ of finite order one has

$$HC_n(RG)_{[g]} \cong \bigoplus_{p+q=n} HC_p(R) \underset{k}{\otimes} H_q(G_g, k).$$

This was first established by Burghelea [8], at least for $R = k$.

In view of Lemma 1.4 which holds in the more general setting of crossed products, it seems worthwhile to try and extend the results of Feigin and Tsygan to crossed products and, in doing so, hopefully also construct more explicit and accessible proofs of these results.

2.4. Hattori-Stallings ranks via cyclic homology. Chern characters provide a factorization of the Hattori-Stallings rank $\chi_S : K_0(S) \to T(S)$ through all even higher cyclic homology groups $HC_{2n}(S)$ (2.2). Therefore, information on the cyclic homology of S has consequences for χ_S. Our aim here is to explain this more specifically for the case of skew group rings $S = R_\alpha G$ by making use of the results of Feigin and Tsygan in 2.3. We will also need the following technical Lemma which elaborates on certain facts in Eckmann [11].

LEMMA. *Let G be a group and suppose that either G is solvable-by-finite or $G \subseteq GL_n(F)$ for some field F of char 0. Then, for all $g \in G$, one has* $\mathrm{hd}_{\mathbb{Q}}(G_g) \leq \mathrm{hd}_{\mathbb{Q}}(G)$.
(Here, $\mathrm{hd}_{\mathbb{Q}}$ denotes homological dimension over $\mathbb{Q}$ and $G_g = \mathbb{C}_G(g)/\langle g \rangle$, as before.)

PROOF: We may clearly assume that $\mathrm{hd}_{\mathbb{Q}}(G) = n < \infty$. The solvable-by-finite case is immediate from a result of Stammbach [39] which states that, for G solvable-by-finite, $\mathrm{hd}_{\mathbb{Q}}(G)$ is equal to the Hirsch number $h(G)$ of G. We can therefore concentrate on the case where G is a linear group over F. Upon replacing G by $\mathbb{C}_G(g)$, we may further assume that $g \in Z :=$ center(G) and $G_g = G/\langle g \rangle$.

By Wehrfritz [40, Theorem 6.2], G/Z is again a linear group over F. Let U be a finitely generated unipotent subgroup of G/Z. Then U, as well as its preimage in G, say V, are nilpotent and so we conclude from Stammbach [39] that

$$n = \mathrm{hd}_{\mathbb{Q}}(G) \geq \mathrm{hd}_{\mathbb{Q}}(V) = h(V) = h(U) + h(Z).$$

Therefore, $h(U) \leq n - h(Z)$, a fixed upper bound independent of U. It now follows from Alperin-Shalen [1] that every finitely generated subgroup $S \leq G/Z$ has finite (co-) homological dimension over $\mathbb{Q}$. Let T denote the preimage of S in G. Then, by Stammbach [39, Lemma 4],

$$n = \mathrm{hd}_{\mathbb{Q}}(G) \geq \mathrm{hd}_{\mathbb{Q}}(T) = \mathrm{hd}_{\mathbb{Q}}(S) + h(Z).$$

So $\mathrm{hd}_{\mathbb{Q}} \leq n - h(Z)$ holds for all finitely generated subgroups $S \leq G/Z$ and, by a direct limit argument, one concludes that

$$\mathrm{hd}_{\mathbb{Q}}(G/Z) \leq n - h(Z)$$

(cf. Bieri [4, Theorem 4.7]).

Now consider $G/\langle g \rangle$. The spectral sequence for the extension $1 \to Z/\langle g \rangle \to G/\langle g \rangle \to G/Z \to 1$ implies that

$$\begin{aligned}
\mathrm{hd}_{\mathbb{Q}}(G/\langle g \rangle) &\leq \mathrm{hd}_{\mathbb{Q}}(G/Z) + \mathrm{hd}_{\mathbb{Q}}(Z/\langle g \rangle) \\
&\leq n - h(Z) + h(Z) \\
&= n,
\end{aligned}$$

as we have claimed. $\square$

We now apply the foregoing to Hattori-Stallings ranks for skew group algebras $S = R_{\alpha}G$ ($\alpha : G \to \mathrm{Aut}_{k\text{-alg}}(R)$).

PROPOSITION. *Let $S = R_{\alpha}G$ be a skew group algebra over a commutative ring k with $\mathbb{Q} \subseteq k$. Assume that R is k-flat with $\dim R = \mathrm{proj.\,dim}_{R^e}(R) < \infty$. Assume further that $\mathrm{hd}_{\mathbb{Q}}(G) < \infty$ and that either G is solvable-by-finite or $G \subseteq GL_n(F)$ for some field F of char 0.*

Then the Hattori-Stallings rank function $\chi_S : K_0(S) \to T(S) = \underset{[g] \in T(G)}{\oplus} H_0(G_g, V_g)$ (1.4) has image supported by the classes $[g]$ of finite order.

PROOF: In both cases, the lemma guarantees that $\mathrm{hd}_k(G_g) \leq \mathrm{hd}_{\mathbb{Q}}(G_g) \leq \mathrm{hd}_{\mathbb{Q}}(G) < \infty$. Therefore, by 2.3(a), Example 2, we have, for g of infinite order,

$$HC_n(S)_{[g]} = 0 \qquad (n > \mathrm{hd}_{\mathbb{Q}}(G) + \dim R).$$

Since χ_S factors through $HC_{2n}(S)$ and since the shift map σ respects the decomposition of $HC_*(S)$, the assertion follows.   □

The above argument, being based on finiteness of homological dimensions, is rather crude. If $S = kG$ is the group algebra and $g \in G$ has infinite order then the $[g]$-component of the shift map,

$$\sigma_{[g]} : HC_n(S)_{[g]} = H_n(G_g, k) \to HC_{n-2}(S)_{[g]} = H_{n-2}(G_g, k),$$

can be shown to be identical with the cap product map

$$\alpha_g \cap (.) : H_n(G_g, k) \to H_{n-2}(G_g, k),$$

where $\alpha_g \in H^2(G_g, \mathbb{Z})$ is the class corresponding to the extension $1 \to \mathbb{Z} = \langle g \rangle \to C_G(g) \to G_g \to 1$ (Burghelea [8]). Using this fact, one can deal with certain groups of infinite homological dimension (and without a.c.c. on cyclic subgroups), but I don't know of any striking general result in this direction.

We now specialize to the case of group rings $S = RG$ of torsion-free groups G. The corollary below shows that, in the situation considered there, the computation of Hattori-Stallings ranks of projective S-modules can be reduced to computing the ranks of suitable projective R-modules. It also shows that, as far as evaluation of χ_S is concerned, projective S-modules look like they are induced from R.

COROLLARY. *Let $S = RG$ be a group ring with R and G as in the proposition. Assume in addition that G is torsion-free. Then, for all finitely generated projective S-modules P,*

$$\chi_S(P) = \chi_S(P/P\omega G \underset{R}{\otimes} S) = T(\mu) \circ \chi_R(P/P\omega G).$$

(Here, ωG denotes the augmentation ideal of RG and $\mu : R \to RG$ is the canonical embedding.)

PROOF: Let $R \overset{\mu}{\hookrightarrow} RG = S \overset{\varepsilon}{\to} R$ be the embedding and the augmentation map. So $\varepsilon \circ \mu = \text{id}_R$. By 1.2 we obtain a commutative diagram

$$
\begin{array}{ccccc}
T(R) & \overset{T(\mu)}{\longrightarrow} & T(S) & \overset{T(\varepsilon)}{\longrightarrow} & T(R) \\
\Big\uparrow{\scriptstyle\chi_R} & & \Big\uparrow{\scriptstyle\chi_S} & & \Big\uparrow{\scriptstyle\chi_R} \\
K_0(R) & \underset{\text{Ind}_R^S}{\longrightarrow} & K_0(S) & \underset{\text{Ind}_S^R}{\longrightarrow} & K_0(R)
\end{array}
$$

with $\mathrm{Ind}_S^R(P) = P \otimes_S R \cong P/P\omega G$. The last equality in the corollary
follows from the left half of the diagram. Writing $T(S) = \underset{[g]\in T(G)}{\oplus} T(S)_{[g]}$
with $T(S)_{[g]} = T(R)$, as in 1.4, we have $\mathrm{Im}\,T(\mu) = T(S)_{[1]}$, and for $t = (t_{[g]})_{[g]\in T(G)} \in T(S)$ one gets $T(\varepsilon)(t) = \sum t_{[g]} \in T(R)$. Thus $T(\mu) \circ T(\varepsilon)$
is the identity on $T(S)_{[1]}$. Since $\chi_S(P) \in T(S)_{[1]}$, by the proposition, we
conclude from the diagram that

$$\chi_S(P) = (T(\mu) \circ T(\varepsilon) \circ \chi_S)(P) = (\chi_S \circ \mathrm{Ind}_R^S \circ \mathrm{Ind}_S^R)(P)$$
$$= \chi_S(P/P\omega G \otimes_R S),$$

which completes the proof. □

2.5. A ring-theoretic application: Idempotents in group rings.
The following corollary is a minor extension of a result of Marciniak [27].
Further results concerning idempotents in group rings can be obtained by
using the Frobenius map on trace groups (in char p) together with valuations (in char 0). Chapter 2 of [31] is a good reference for these techniques.

COROLLARY. *Let k be a commutative ring with $\mathbb{Q} \subseteq k$ and let R be a k-
algebra that is finitely generated projective over k. Let G be a torsion-free
group as in 2.4. Then, if R has no idempotents $\neq 0, 1$, neither does RG.*

PROOF: Let $e \in RG$ be an idempotent. Then $P = eRG$ is finitely generated projective over the group algebra kG. Therefore, the corollary in 2.4
gives $\chi_{kG}(P) = \chi_{kG}(P/P\omega G \otimes_k kG)$. But $P/P\omega G \otimes_k kG \cong \varepsilon(e)RG$, where
$\varepsilon : RG \to R$ is the augmentation map. By assumption on R, either $\varepsilon(e) = 0$
or $\varepsilon(e) = 1$. Putting $e_1 = e$ in the former case and $e_1 = 1 - e$ in the latter,
we see that $P_1 = e_1 RG$ is a finitely generated projective kG-module with
$\chi_{kG}(P_1) = 0$. A well-known theorem of Kaplansky (see [31, Exercise 9 on
p. 65]) implies that $P_1 = 0$, which proves the corollary. □

§3. Grothendieck groups: Moody's theorem and applications

3.1. This section is entirely devoted to a discussion of Moody's induction
theorem for crossed products $R * G$ of polycyclic-by-finite groups G [29],
[30]. We give some background information in 3.3, a very rough outline of
the proof in 3.4, and two recent applications in 3.5 and 3.6. The applications
are the solution of the "Goldie rank conjecture" for prime polycyclic group
rings and the computation of the rank of $G_0(kG)$ extending classical results
for finite groups due to Brauer and Frobenius.

The following **notation** will be kept throughout this section.

$S = R * G$ will be a crossed product with

R a right Noetherian ring, and

G a polycyclic-by-finite group.

In particular, S is a right Noetherian ring. We also put

$$\mathcal{F}_{(\mathrm{max})} = \mathcal{F}_{(\mathrm{max})}(G) = \{ \text{ all (maximal) finite subgroups of } G\}.$$

Then G acts by conjugation on $\mathcal{F}$ and $\mathcal{F}_{\mathrm{max}}$, and we let $\mathcal{F}_{(\mathrm{max})}/G$ denote a full representative set of non-conjugate (maximal) finite subgroups of G. It is known that $\mathcal{F}_{(\mathrm{max})}/G$ has only finitely many members (e.g., Segal [**35**, Theorem 5 on p. 175]).

3.2. Moody's induction theorem ([**29**], [**30**]): *The map*

$$\underset{X \in \mathcal{F}_{\mathrm{max}}/G}{\oplus} \mathrm{Ind}_{R*X}^{S} : \underset{X \in \mathcal{F}_{\mathrm{max}}/G}{\oplus} G_0(R * X) \to G_0(S)$$

is surjective.

Here, $\mathrm{Ind}_{R*X}^{S} = (.) \underset{R*X}{\otimes} S$ is the usual induction map that is given by the embedding $R * X \subseteq R * G = S$.

3.3. Some predecessors. The earliest forerunner of Moody's theorem is Grothendieck's theorem which deals with the case $G = \mathbb{Z}$. It was proved in the late 50's.

GROTHENDIECK'S THEOREM. *Let R be an algebra over a commutative Noetherian ring k and assume that R is finitely generated as k-module. Put $T = R[t^{(\pm 1)}]$, the polynomial ring or Laurent polynomial ring over R. Then $\mathrm{Ind}_R^T = (.) \underset{R}{\otimes} T : G_0(R) \to G_0(T)$ is an isomorphism.*

For a proof, see Bass [**2**, p. 640]. The essential point is to prove surjectivity of Ind_R^T. The polynomial ring case of Grothendieck's theorem is greatly generalized in Quillen's theorem.

QUILLEN'S THEOREM ([**33**]). *Let $T = \underset{i \geq 0}{\oplus} T_i$ be a $\mathbb{Z}^+$-graded right Noetherian ring and put $R = T_0$. Assume that ${}_R T$ and ${}_T R = {}_T(T/\underset{i>0}{\oplus} T_i)$ both have finite weak (= Tor) dimension. Then*

$$\mathrm{Ind}_R^T = \sum_{i \geq 0} \mathrm{Tor}_i^R(., T) : G_0(R) \to G_0(T)$$

is an isomorphism.

Again, surjectivity is the crux here whereas a back map $G_0(T) \to G_0(R)$ is easy to construct. In most applications $_RT$ is projective, or at least flat, in which case $\mathrm{Ind}_R^T = (.) \underset{R}{\otimes} T$, as usual. An algebraic proof of Quillen's theorem is given in Section 34 of Passman's book [**32**] (based on the paper [**10**] by Cliff and Weiss).

Later, special cases and weaker forms of Moody's theorem have been established by Farrell and Hsiang [**14**], [**15**], Brown, Howie, and Lorenz [**6**], and by Quinn [**34**].

3.4. Steps in the proof of Moody's theorem. Moody's original proof has been considerably reworked by Cliff and Weiss [**10**], and by Farkas and Linnell [**13**]. A combination of these proofs is presented in [**32**]. We list the main steps in this proof and give an outline of the proof of Step 1.

First, by an easy inductive argument, one reduces to the case where G is finitely generated abelian-by-finite. So G has a finitely generated abelian normal subgroup A with $\overline{G} = G/A$ finite. Conjugation makes A a finitely generated module over the integral group ring $\mathbb{Z}\overline{G}$. The proof proceeds in three steps of increasing generality.

Step 1. A is free as $\mathbb{Z}\overline{G}$-module.

Step 2. $A \underset{\mathbb{Z}}{\otimes} \mathbb{Q}$ is free as $\mathbb{Q}\overline{G}$-module.

Step 3. $A \underset{\mathbb{Z}}{\otimes} \mathbb{Q}$ is projective as $\mathbb{Q}\overline{G}$-module ($=$ general case, by Maschke's theorem).

Step 2 is technically the most involved. Step 1 is a nice application of Quillen's theorem. The argument is as follows.

If A is free over $\mathbb{Z}\overline{G}$ then $G \cong A \rtimes \overline{G}$, a semidirect product, because $H^2(\overline{G}, A) = \{0\}$. Fix a $\mathbb{Z}$-basis $\mathcal{B}$ of A that is permuted by G-conjugation and put

$$X_i = \{\text{products of length } i \text{ with factors } \in \mathcal{B}\}, \text{ and}$$
$$X = \underset{i \geq 0}{\cup} X_i \subseteq A.$$

Then $\overline{G}X_i = X_i\overline{G}$ holds for all i. Thus $T := R * X\overline{G}$ is a subring of $S = R * G$, and T is graded by $T_i = R * X_i\overline{G}$ ($i \geq 0$).

The hypotheses of Quillen's theorem are easily checked, and $\mathrm{Ind}_{T_0}^T = (.) \underset{T_0}{\otimes} T : G_0(T_0) \to G_0(T)$ is surjective (indeed, an isomorphism). Note that $T_0 = R * \overline{G}$ and that $S = R * G$ is a classical localization of T. Therefore, $\mathrm{Ind}_T^S : G_0(T) \to G_0(S)$ is epi and, combining the two induction maps, we obtain an epimorphism $\mathrm{Ind}_{R*\overline{G}}^S = (.) \underset{R*\overline{G}}{\otimes} S : G_0(R * \overline{G}) \twoheadrightarrow G_0(S)$. This finishes the proof of Step 1. $\qquad\square$

3.5. Application: Goldie ranks.

Recall that (for any right Noetherian ring S), Goldie's reduced rank function is a homomorphism

$$\rho = \rho_S : G_0(S) \to \mathbb{Z}$$

which can be defined as follows. Let N be the nilpotent radical of S and put $Q = \mathrm{Fract}(S/N)$, the classical semisimple Artinian ring of fractions of S/N. Then ρ_S is the composite map

$$\rho_S : G_0(S) \xrightarrow{\;\cong\;} G_0(S/N) \xrightarrow{\;\mathrm{Ind}_{S/N}^Q\;} G_0(Q) \xrightarrow{\text{composition length over } Q} \mathbb{Z}$$

$$[V] \longmapsto \sum_{i \geq 0} [VN^i/VN^{i+1}]$$

For example, if S is prime then $Q = \mathrm{Fract}(S) \cong M_n(D)$ for some division ring D, and $\rho(S) = n$.

Following [21] and [32] we now describe how Moody's theorem can be used to determine $\rho(S)$ for $S = R * G$ as in 3.1, provided S is prime.

DEFINITION: Given $R * X$ with X a finite group. Consider the homomorphism

$$\varphi : G_0(R * X) \xrightarrow{\mathrm{Res}_R^{R*X}} G_0(R) \xrightarrow{\rho_R} \mathbb{Z}.$$

Let $m(R * X) \in \mathbb{Z}$ be the positive generator of $\mathrm{Im}\,\varphi$ and define

$$\mathrm{index}(R * X) = \varphi(R * X)/m(R * X) = \rho_R(R) \cdot |X|/m(R * X) \in \mathbb{Z}_{>0}.$$

Examples and Remarks. (1) If $R * X = RX$ is the group ring then φ is epi, as Res_R^{RX} is. So

$$\mathrm{index}(RX) = \rho_R(R) \cdot |X|.$$

(2) Note that $\varphi = \varphi_R$ can be factored as

$$\varphi : G_0(R * X) \xrightarrow{\cong} G_0((R/N) * X) \xrightarrow{\text{Ind}} G_0(Q * X) \xrightarrow{\varphi_Q} \mathbb{Z},$$

where N is the nilpotent radical of R, $Q = \text{Fract}(R/N)$, $(R/N) * X \cong (R * X)/(N * X)$, and $Q * X \cong \text{Fract}((R/N) * X)$. It follows that

$$m(R * X) = m(Q * X) = \gcd_V \quad \{\text{composition length of } V \text{ over } Q\},$$

where V runs over the irreducible $Q * X$-modules. Moreover,

$$\text{index}(R * X) = \frac{\rho_R(R)}{\rho_Q(Q)} \cdot \text{index}(Q * X).$$

(3) Suppose that R is semiprime. Then $R * X$ is a domain exactly if $\text{index}(R * X) = 1$. To see this, note that if $R * X$ is a domain then $\rho_R(R) = 1$ and $Q * X$ is a division ring. Thus $m(Q * X) = |X|$ and hence $\text{index}(R * X) = 1$. Conversely, if $\text{index}(R * X) = 1$ then $\text{index}(Q * X) = 1$ and so $m(Q * X) = |X| \cdot \rho_Q(Q) = $ composition length of $Q * X$ over Q. This shows that $Q * X$ is a division ring, and hence $R * X$ is a domain.

THEOREM ([21], [32]). *Let $S = R * G$ be as in 3.1 and assume that S is prime. Then $\rho(S) = \underset{X \in \mathcal{F}_{\max}/G}{\text{lcm}} \{\text{index}(R * X)\}.$*

We note three special cases of particular importance. In each case, our assumptions imply that S is prime (see [32, Proposition 8.3]).

COROLLARY. *Let $S = R * G$ be as in 3.1.*

   i. *S is a domain if and only if $R * X$ is a domain for all $X \in \mathcal{F}_{\max}/G$.*

   ii. *If R is G-prime and G is torsion-free then S is prime and $\rho(S) = \rho(R)$.*

   iii. *Suppose that R is prime and G has no finite normal subgroups $\neq \langle 1 \rangle$. Then the group ring RG is prime and $\rho(RG) = \rho(R) \cdot \underset{X \in \mathcal{F}_{\max}/G}{\text{lcm}} \{|X|\}.$*

Part (iii), due to Moody [29], constitutes the affirmative solution of the so-called "Goldie rank conjecture" of Farkas [12] and Rosset [35].

Outline of the proof of the theorem. The normalized reduced rank function is defined by

$$\rho_S' : G_0(S) \to \mathbb{Q}$$
$$\cup \qquad \cup$$
$$[V] \mapsto \rho_S(V)/\rho_S(S)$$

Using the fact that S is prime one shows [21] that, for all $X \in \mathcal{F}$ and each finitely generated $R * X$-module W,

$$\rho'_S(W \underset{R*X}{\otimes} S) = |X|^{-1}\rho'_R(W) = \frac{\rho_R(W)}{|X|\rho_R(R)}.$$

Therefore, $\rho'_S(\operatorname{Ind}_{R*X}^S G_0(R*X)) = \operatorname{index}(R*X)^{-1} \cdot \mathbb{Z}$, and Moody's theorem 3.2 implies that

$$\frac{1}{\rho(S)}\mathbb{Z} = \operatorname{Im}\rho'_S = \sum_{X \in \mathcal{F}_{\max}/G} \rho'_S(\operatorname{Ind}_{R*X}^S G_0(R*X))$$

$$= \sum_{X \in \mathcal{F}_{\max}/G} \operatorname{index}(R*X)^{-1} \cdot \mathbb{Z},$$

which gives the equality in the theorem.   □

3.6. Application: A Brauer-Frobenius theorem for polycyclic-by-finite groups. We now specialize to the situation where

$S = kG$ is the group algebra of the polycyclic-by-finite
group G over the field k.

We also assume, for simplicity, that

k is a splitting field for all $X \in \mathcal{F} = \mathcal{F}(G)$.

For the general case we refer to [24]. Finally, we put

$p = \operatorname{char} k \ (\geq 0)$, and
$T(G)_{p'} = \{[g] \in T(G) \mid g \text{ has finite order, not divisible by } p\}$,

the set of *p-regular torsion conjugacy classes* of G. (For $p = 0$ the non-divisibility condition is of course vacuous.)

Since each of the groups $G_0(kX)$, for $X \in \mathcal{F}_{\max}$, is free abelian of finite rank and since $\mathcal{F}_{\max}/G$ is finite, Moody's theorem 3.2 implies in particular that $G_0(kG)$ is a finitely generated abelian group. The following result determines the rank of this group.

THEOREM ([24]). $\operatorname{rank} G_0(kG) = |T(G)_{p'}|$.

Before we describe an outline of the proof, we remark that the case where G is finite is classical. Indeed, by results of Frobenius and Brauer, one knows that for *finite* G

$G_0(kG) \cong \mathbb{Z}^r$ with $r = \#$ isomorphism classes of simple kG-modules
$= |T(G)_{p'}|$.

For general polycyclic-by-finite groups G, $G_0(kG)$ need not be torsion-free anymore. We illustrate the theorem by a simple example; more involved examples can be found in [25].

Example: Let $G = \langle x, y \mid y^2 = 1, xy = yx^{-1}\rangle$ be the infinite dihedral group. Then G has three conjugacy classes of finite order, namely $[1], [y], [xy]$, with corresponding orders 1,2,2. Thus the theorem implies that

$$\operatorname{rank} G_0(kG) = \begin{cases} 1 & \text{if } p = 2 \\ 3 & \text{otherwise.} \end{cases}$$

We further remark that Moody's theorem gives the following explicit generators for $G_0(kG)$: $\alpha_1 = \operatorname{Ind}_{\langle 1\rangle}^G[k] = [kG]$, $\alpha_2 = \operatorname{Ind}_{\langle y\rangle}^G[k]$, and $\alpha_3 = \operatorname{Ind}_{\langle xy\rangle}^G[k]$. In case $p \neq 2$ we deduce from the rank of $G_0(kG)$ that $\alpha_1, \alpha_2, \alpha_3$ must be free generators, and hence $G_0(kG) \cong \mathbb{Z}^3$. If $p = 2$ then $2\alpha_2 = 2\alpha_3 = \alpha_1$, and so α_2 and $\beta = \alpha_2 - \alpha_3$ suffice to generate $G_0(kG)$. It can be shown [25] that $\beta = [k]$ has order 2. So $G_0(kG) \cong \mathbb{Z} \oplus \mathbb{Z}/2\mathbb{Z}$ holds in this case.

We finish with a rough *outline of the proof of the theorem*. For more details see [24].

If $p > 0$, fix a p-modular system (K, R, k) with R a complete d.v.r. of characteristic 0, k the residue field of R, and K the field of fractions of R. In case $p = 0$ take $K = k$ so that, in either case, char $K = 0$.

The set $\mathcal{F} = \mathcal{F}(G)$ becomes a category, the so-called *Frobenius category* of G, with morphisms being given by inclusions of conjugates. So, if $X_1^g \subseteq X_2$ for $X_1, X_2 \in \mathcal{F}$ and $g \in G$, then one has an $\mathcal{F}$-morphism $X_1 \to X_2, x_1 \mapsto x_1^g$. The assignments

$$F : X \mapsto K^{T(X)_{p'}} \text{ and } G : X \mapsto G_0(kX) \underset{\mathbb{Z}}{\otimes} K$$

define functors F and G from $\mathcal{F}$ to the category of K-vector spaces. Now, for each finite group X, one has an isomorphism

$$\operatorname{ch}_X : G_0(kX) \underset{\mathbb{Z}}{\otimes} K \xrightarrow{\cong} K^{T(X)_{p'}}$$

given by $\operatorname{ch}_X(V)(x) = |\mathbb{C}_X(x)|^{-1} \cdot \lambda_V(x)$ ($x \in X$ a p'-element), where λ_V is the Brauer or ordinary character of the kX-module V. Moreover these isomorphisms yield a natural equivalence between F and G. (We take this opportunity to point out that, in the original paper [24], we neglected to include the factor $|\mathbb{C}_X(x)|^{-1}$ in the definition of ch_X. It is needed in order to make the appropriate diagrams commute, as it is required for a natural transformation of functors.)

We now consider the colimits of F and G over the category $\mathcal{F}$. One easily shows that

$$\varinjlim_{X \in \mathcal{F}} K^{T(X)_{p'}} \cong K^{T(G)_{p'}},$$

and Moody's theorem 3.2 furnishes us with at least an epimorphism

$$\Phi : \varinjlim_{X \in \mathcal{F}} G_0(kX) \underset{\mathbb{Z}}{\otimes} K \twoheadrightarrow G_0(kG) \underset{\mathbb{Z}}{\otimes} K.$$

The last ingredient needed is the existence of a finite image $\overline{G} = G/N$ of G, with N torsion-free, so that the canonical map $G \to \overline{G}$ yields an *injection* $T(G)_{p'} \hookrightarrow T(\overline{G})_{p'}$ [24, Lemma 1]. Combining all this, we obtain the following commutative diagram ($\varinjlim$ stands for $\varinjlim_{X \in \mathcal{F}}$).

$$
\begin{array}{ccccc}
\varinjlim K^{T(X)_{p'}} & \xrightarrow{\ \cong\ } & K^{T(G)_{p'}} & \hookleftarrow & K^{T(\overline{G})_{p'}} \\
{\scriptstyle\cong}\big\uparrow{\scriptstyle\varinjlim \mathrm{ch}_X} & & & & {\scriptstyle\cong}\big\uparrow{\scriptstyle\mathrm{ch}_{\overline{G}}} \\
\varinjlim G_0(kX) \otimes K & \xrightarrow[\text{via } X \cong \bar{X} \subseteq \bar{G}]{\text{induction}} & G_0(k\overline{G}) \otimes K & \ni & \sum (-1)^i [H_i(N,V)] \\
& {\scriptstyle \Phi}\searrow \qquad \nearrow & & \nearrow & \\
& G_0(kG) \otimes K & \ni & [V] &
\end{array}
$$

We conclude that Φ must be injective also, and hence

$$G_0(kG) \underset{\mathbb{Z}}{\otimes} K \cong \varinjlim_{X \in \mathcal{F}} G_0(kX) \underset{\mathbb{Z}}{\otimes} K \cong K^{T(G)_{p'}},$$

which implies the theorem. $\square$

REFERENCES

1. R.C. Alperin and P.B. Shalen, *Linear groups of finite cohomological dimension*, Bull. Amer. Math. Soc. (new series) **4** (1981), 339–341.
2. H. Bass, "Algebraic K-Theory," Benjamin, New York, 1968.
3. H. Bass, *Euler characteristics and characters of discrete groups*, Invent. Math. **35** (1976), 155–196.
4. R. Bieri, "Homological dimension of discrete groups," 2nd ed., Queen Mary College Math. Notes, University of London, 1981.
5. N. Bourbaki, "Algèbre," chap. 1–3, Hermann, Paris, 1970.
6. K.A. Brown, J. Howie, and M. Lorenz, *Induced resolutions and Grothendieck groups of polycyclic-by-finite groups*, J. Pure and Appl. Algebra **53** (1988), 1–14.
7. K.S. Brown, "Cohomology of groups," Springer, New York, 1982.
8. D. Burghelea, *The cyclic homology of group rings*, Comment. Math. Helvetici **60** (1985), 354–365.
9. H. Cartan and S. Eilenberg, "Homological algebra," Princeton University Press, Princeton, 1956.
10. G.H. Cliff and A. Weiss, *Moody's induction theorem*, Illinois J. Math. **32** (1988), 489–500.
11. B. Eckmann, *Cyclic homology of groups and the Bass conjecture*, Comment. Math. Helvetici **61** (1986), 193–202.
12. D.R. Farkas, *Group rings: an annotated questionnaire*, Comm. Algebra **8** (1980), 585–602.
13. D.R. Farkas and P.A. Linnell, *Zero divisors in group rings: something old, something new*, Contemporary Math. **93** (1989), 155–166.
14. F.T Farrell and W.C. Hsiang, *The topological Euclidean space form problem*, Invent. Math. **45** (1978), 181–192.
15. F.T Farrell and W.C. Hsiang, *The Whitehead group of poly-(finite or cyclic) groups*, J. London Math. Soc. (2) **24** (1981), 308–324.
16. B.L. Feigin and B.L. Tsygan, *The cyclic homology of algebras with quadratic relations, universal enveloping algebras and group algebras*, In: Lect. Notes in Math., Vol. 1289, 210–239, Springer, Berlin. 1987
17. A. Hattori, *The rank element of a projective module*, Nagoya J. Math. **25** (1965), 113–120.
18. T.J. Hodges, *K-theory of Noetherian rings*, in "Seminaire d'algèbre," (M.P. Malliavin, ed.), Lect. Notes in Math., Springer, Berlin, 1990, pp. 246–269.
19. M. Karoubi, "Homologie cyclique et K-théorie," Astérisque No. 149, Soc. Math. de France, 1987.
20. J.-L. Loday and D. Quillen, *Cyclic homology and the Lie algebra homology of matrices*, Comment. Math. Helvetici **59** (1984), 565–591.
21. M. Lorenz, *Goldie ranks of prime polycyclic crossed products*, in "'Perspectives in Ring Theory," (F. van Oystaeyen and L. Le Bruyn, eds.), Kluwer, Dordrecht, 1988, pp. 215–219.
22. M. Lorenz, K_0 *of skew group rings and simple Noetherian rings without idempotents*, J. London Math. Soc. (2) **32** (1985), 41–50.
23. M. Lorenz, *Frobenius reciprocity and* G_0 *of skew group rings*, in "Ring theory," (J.L Bueso, P. Jara, B. Torrecillas, eds.), Lect. Notes in Math., Vol. 1328, Springer, Berlin, 1988, pp. 165–172.
24. M. Lorenz, *The rank of* G_0 *for polycyclic group algebras*, Banach Center Publications, Vol. 26, Warsaw. (to appear)
25. M. Lorenz and D.S. Passman, *The structure of* G_0 *for certain polycyclic group algebras and related algebras*, Contemporary Math. **93** (1989), 283–302.

26. S. Mac Lane, "Homology," Springer, Berlin, 1975.
27. Z. Marciniak, *Cyclic homology and idempotents in group rings*, Inst. of Mathematics, Warsaw University. preprint
28. S. Montgomery, "Fixed rings of finite automorphism groups of associative rings," Lect. Notes in Math., Vol. 818, Springer, Berlin, 1980.
29. J.A. Moody, *Induction theorems for infinite groups*, PhD. thesis (1986), Columbia University.
30. J.A. Moody, *Brauer induction for G_0 of certain infinite groups*, J. Algebra **122** (1989), 1–14.
31. D.S. Passman, "The algebraic structure of group rings," Wiley-Interscience, New York, 1977.
32. D.S. Passman, "Infinite crossed products," Academic Press, San Diego, 1989.
33. D. Quillen, *Higher algebraic K-theory I*, in "Algebraic K-theory I," Lect. Notes in Math., Vol. 341, Springer, Berlin, 1973, pp. 77–139.
34. F. Quinn, *Algebraic K-theory of poly-(finite or cyclic) groups*, Bull. Amer. Math. Soc. **12** (1985), 221–226.
35. S. Rosset, *The Goldie rank of virtually polycyclic group rings*, in "The Brauer group," Lect. Notes in Math., Vol. 844, Springer, Berlin, 1981, pp. 35–45.
36. D. Segal, "Polycyclic groups," Cambridge University Press, Cambridge, 1983.
37. P. Seibt, "Cyclic homology of algebras," World Scientific, Singapore, 1987.
38. J.R. Stallings, *Centerless groups—an algebraic formulation of Gottlieb's theorem*, Topology **4** (1965), 129–134.
39. U. Stammbach, *On the weak homological dimension of the group algebra of solvable groups*, J. London Math. Soc. (2) **2** (1970), 567–570.
40. B.A.F. Wehrfritz, "Infinite linear groups," Springer, Berlin, 1973.

Department of Mathematics
Temple University
Philadelphia, PA 19122

Research supported in part by a grant from the National Science Foundation.

Prime Ideals in Crossed Products

D. S. PASSMAN

Abstract. This talk is concerned with crossed products and, to a lesser extent, with more general group-graded rings. Specifically, we consider when such rings are prime or semiprime and we study the nature of their prime ideals under certain finiteness assumptions. We also include some applications to the Galois theory of noncommutative rings.

§1. Introduction

I started my career as a group theorist and then slowly moved into ring theory. I was always interested in the interplay between these two subjects, and there are at least two obvious places to look. First, we have the Galois theory of rings. Here, if R is any ring, then $\mathrm{Aut}(R)$ is a group, the group of all automorphisms of R. In particular, if G is any subgroup of $\mathrm{Aut}(R)$, then one can study the relationship between R and the fixed ring R^G, as well as between intermediate rings and subgroups of G. We will consider this later in the talk. Second, for any ring R, $\mathrm{U}(R)$ is also a group, the group of units of R. For example, if R is the matrix algebra $\mathrm{M}_n(K)$, then $\mathrm{U}(R) = \mathrm{GL}_n(K)$, the general linear group over the field K. One can of course study subgroups G of $\mathrm{U}(R)$ and this leads us to consider group rings $R = S[G]$.

Suppose that R is a K algebra and that G is given. Then the group algebra $K[G]$ is actually a universal object for this situation. Specifically, any group homomorphism $G \to \mathrm{U}(R)$ extends to an algebra homomorphism $K[G] \to R$. Of course, if R is just a ring, then we can view it as a Z-algebra and obtain a homomorphism $Z[G] \to R$ from the integral group ring $Z[G]$. Again let R be a K-algebra and say $G \subseteq \mathrm{U}(R)$. Then without loss, we can assume that R is generated by G and K, so that the map $\xi\colon K[G] \to R$ is an epimorphism. It also makes sense to assume that R has some reasonable ring theoretic structure. Thus for example, if R is a prime ring, then the kernel P of ξ is a prime ideal of $K[G]$ and an understanding of P yields information on R. We will get to this in due time.

The study of group algebras originated as part of the representation theory of finite groups. Thus suppose G is such a group and that K is a field. Frobenius considered representations in an entirely character theoretic manner and Schur viewed them as homomorphisms $G \to \mathrm{GL}_n(K)$. It remained for E. Noether to observe the homomorphism $K[G] \to \mathrm{M}_n(K)$. A classical theorem of Maschke (1898) asserts that $K[G]$ is semiprime if and only if $|G| \neq 0$ in K. Since $\dim_K K[G] < \infty$, semiprime is equivalent to semiprimitive and in fact to all $K[G]$-modules being completely reducible. In particular, if K is the field of complex numbers, then $K[G] = \oplus \sum_i \mathrm{M}_{n_i}(K)$. Furthermore, we have

LEMMA 1.1. *Let G be a finite group.*

(i) *$\mathbf{Z}(K[G])$, the center of $K[G]$, has as a K-basis the set of conjugacy class sums of G.*

(ii) *If K is the complex field, then the number of irreducible representations of $K[G]$ is equal to the number of conjugacy classes of G.*

It is surprising how many properties of $K[G]$ relate to the center of this algebra. Before we consider the prime ideals of this ring, it is appropriate to first mention certain more general structures.

We say that S is a *G-graded ring* if S is the direct sum $S = \oplus \sum_{x \in G} S_x$ of the additive subgroups S_x, indexed by the elements $x \in G$, and if $S_x S_y \subseteq S_{xy}$ for all $x, y \in G$. It is easy to see that $1 \in S_1$ and hence that S_1 is a subring of S. For example, the group algebra $S = K[G]$ is G-graded by defining $S_x = Kx$. Unlike group rings, there is so much slack in this definition that the presence of the group can become somewhat blurred. For example, if S is H-graded and if H is a subgroup of any group G, then S is G-graded by defining $S_x = 0$ for all $x \in G \setminus H$. Furthermore, suppose S is H_1-graded and T is H_2-graded. Then $S \oplus T$ becomes G-graded for any group G containing disjoint copies of H_1 and H_2. Specifically, we set

$$(S \oplus T)_x = \begin{cases} S_1 \oplus T_1, & \text{if } x = 1 \\ S_x, & \text{if } x \in H_1 \setminus 1 \\ T_x, & \text{if } x \in H_2 \setminus 1 \\ 0, & \text{otherwise.} \end{cases}$$

Thus without some nontriviality assumption, a group-graded ring can become a "thing-graded ring". Some nontriviality candidates for a G-graded ring S are as follows.

(1) We could merely assume that $S_x S_y \neq 0$ for all $x, y \in G$.

(2) We say that S is *strongly graded* if $S_x S_y = S_{xy}$ for all $x, y \in G$. It is easy to see that this is equivalent to $S_x S_{x^{-1}} = S_1$ for all $x \in G$ and then to $1 \in S_x S_{x^{-1}}$.

(3) We could assume that each component S_x contains a unit, say $\bar{x}$. As we quickly see, this forces a rather tight structure on S. Indeed,

let $R = S_1$ and notice that $\bar{x}^{-1} \in S_{x^{-1}}$. Thus $S_x \bar{x}^{-1} \subseteq S_1 = R$ and it follows that $S_x = R\bar{x}$ and $S = \oplus \sum_{x \in G} R\bar{x}$. Next, since $\bar{x}\bar{y}$ is a unit in $S_{xy} = R\overline{xy}$, we have $\bar{x}\bar{y} = u_{x,y}\overline{xy}$ for some $u_{x,y} \in U(R)$. Furthermore, $\bar{x}^{-1}R\bar{x} = R$ and thus conjugation by $\bar{x}$ induces an automorphism of R. Rings of this nature are called *crossed products* and we write $S = R*G$. We rarely construct crossed products; they occur naturally. Here are some examples.

a. Suppose H is a group, $N \lhd H$ is a normal subgroup and $G = H/N$. Just as H is somehow determined by N and H/N, we might expect $S = K[H]$ to be somehow determined by its subring $R = K[N]$ and the quotient group $H/N = G$. This is indeed the case and, to start with, we observe that S is G-graded by defining $S_g = \sum_{h \in N\bar{g}} Kh$ for all $g \in G$. Here, $\bar{g} \in H$ is any inverse image of g under the epimorphism $H \to H/N = G$, so that the coset $N\bar{g}$ is the set of all such inverse images. Note that $S_1 = R$ and that S_g contains the unit $\bar{g}$. Thus $K[H] = S = R*G = K[N]*(H/N)$. Similarly, we have

LEMMA 1.2. *If $S = R*H$ and $N \lhd H$, then $S = (R*N)*(H/N)$.*

Thus crossed products have an important closure property not enjoyed by group algebras. This frequently allows for inductive proofs which really have no group algebra analog.

b. Crossed products first occurred in the classical structure theorems for finite dimensional division rings. Thus suppose D is a division ring, finite dimensional over its center K, and let F be a maximal subfield of D. It is known that such an F exists with F/K separable, and let us assume further that F/K is normal with $G = \text{Gal}(F/K)$. If $g \in G$ then, by the Skolem-Noether Theorem, the automorphism $g \colon F \to F$ can be realized as the restriction of the inner automorphism of D induced by some $\bar{g} \in D$. One can then show that $\bar{G} = \{\bar{g} \mid g \in G\}$ is F-linearly independent and then, by dimension considerations, that D is generated by F and $\bar{G}$. Thus $D = \oplus \sum_{g \in G} F\bar{g} = F*G$. More generally, we have

THEOREM 1.3. (Brauer). *Let A be a central simple algebra with $\mathbb{Z}(A) = K$. Then there exist n, F and G with $M_n(A) = F*G$.*

This leads rather quickly to the characterization of the Brauer group of K as a suitable second cohomology group.

c. Suppose G acts as automorphisms on R. By this we mean that there is a group homomorphism $G \to \text{Aut}(R)$ given by $g \mapsto {}^g$. Then we can form the *skew group ring* RG which is a somewhat simpler version of a crossed product. Here RG has the elements of G as a left R-basis and multiplication is determined distributively by $rg \cdot sh = (rs^{g^{-1}})(gh)$ for all $r, s \in R$ and $g, h \in G$. It is easy to verify that RG is an associative ring which clearly contains all the ingredients of this Galois theoretic situation. Thus crossed products results should translate to results on group actions and we will see an example of this later on.

d. Finally suppose R is a domain with center K and let G be a torsion free nilpotent group of units of R. Assume that R is generated by G and K, so that clearly $\mathbb{Z}(G) \subseteq \mathbb{Z}(R) = K$. Then it can be shown that coset representatives for $\mathbb{Z}(G)$ in G form a K-basis for R and therefore we have

THEOREM 1.4 (Zalesskiĭ). *In the above situation, $R = K*\big(G/\mathbb{Z}(G)\big)$.*

Let us see what this says about the epimorphism $\xi \colon K[G] \to R$. To this end, write $W = \mathbb{Z}(G)$ and think of G/W as a set of coset representatives for W in G. Then every element $\alpha \in K[G]$ can be written uniquely as a finite sum $\alpha = \sum_{x \in G/W} \alpha_x x$ with $\alpha_x \in K[W]$. If $I = \mathrm{Ker}(\xi)$, then it follows that $\alpha \in I$ if and only if $\alpha_x \in I$ for all $x \in G/W$. In other words, $I = (I \cap K[W])K[G]$. Conversely,

LEMMA 1.5. *Let $H \lhd G$ and $I \lhd K[G]$. If $I = (I \cap K[H])K[G]$, then $K[G]/I \cong R*(G/H)$ where $R = K[H]/(I \cap K[H])$.*

In the above situation we say that H *controls* I and one can show that every ideal I has a unique minimal controller subgroup. There are numerous group ring theorems which assert that certain ideals I are controlled by certain subgroups. In particular, these results really assert that $K[G]/I$ is an appropriate crossed product.

§2. Prime and semiprime rings

If we wish to understand the prime and semiprime ideals of a ring, we should first try to decide whether 0 is such an ideal. That is, we should first determine whether the ring itself is a prime or semiprime ring. For group algebras, these are old results proved using Δ-*methods*. While we will not describe these methods here, we will at least define Δ.

Recall that for G a finite group, $\mathbb{Z}(K[G])$ has a K-basis consisting of the class sums of G. This is, in fact, also true for G infinite provided that we restrict our attention to those conjugacy classes of finite size. Since the size of the class of x is $|G : \mathbb{C}_G(x)|$, we define

$$\Delta(G) = \{\, x \in G \mid |G : \mathbb{C}_G(x)| < \infty \,\}$$

and

$$\Delta^+(G) = \{\, x \in \Delta(G) \mid o(x) < \infty \,\}$$

where $o(x)$ is the order of x. Furthermore, let

$$\Delta^p(G) = \langle\, x \in \Delta(G) \mid x \text{ is a } p\text{-element} \,\rangle$$

for all primes p. Basic properties are as follows.

LEMMA 2.1. *Let G be a group.*

(i) *Δ is a characteristic subgroup of G called the f.c. center.*
(ii) *Δ^+ is generated by all the finite normal subgroups of G.*
(iii) *Δ/Δ^+ is torsion free abelian.*

We can now state the group algebra results and we have chosen to formulate them for easy comparison with their crossed product generalizations. The proposition below is classical.

THEOREM 2.2. (Connell). *The following are equivalent.*

(i) *$K[G]$ is prime.*
(ii) *G has no nonidentity finite normal subgroup.*
(iii) *$\Delta^+(G) = 1$.*

PROPOSITION 2.3. *If K has characteristic 0, then $K[G]$ is semiprime.*

THEOREM 2.4. *If K has characteristic $p > 0$, then the following are equivalent.*

(i) *$K[G]$ is semiprime.*
(ii) *G has no finite normal subgroup of order divisible by p.*
(iii) *$\Delta^p(G) = 1$.*

Now for crossed products. If $R*G$ is given, then each element of G acts on R and this determines a permutation action of G on the ideals of R. We say that R is *G-prime* if for all nonzero G-stable ideals A, B we have $AB \neq 0$. Similarly, R is *G-semiprime* if for all nonzero G-stable ideals A we have $A^2 \neq 0$. When $G = 1$, this of course reduces to the usual definitions of prime and semiprime. These concepts are relevant, since

LEMMA 2.5. *If $R*G$ is prime (or semiprime), then R is G-prime (or G-semiprime).*

Now it turns out that the Δ-methods are actually quite difficult to apply here. This is as it should be since the crossed product results are not quite the same as those for group algebras. Primeness and semiprimeness do depend on the finite subgroups N of G, but these groups N need not be normal. Indeed we can measure the failure of normality by considering $H = \mathbb{N}_G(N)$. Obviously, anything we can say about H will help our understanding of this situation and there are two key claims which can be made.

1. If $I \triangleleft R$, let G_I denote the stabilizer of I in G. The first claim is that $H = \mathbb{N}_G(N)$ must equal G_I for some I. Thus for example if the crossed product is an ordinary group ring $R[G]$, then the action is trivial so $H = G_I = G$. In other words, in this case it is still the finite normal subgroups that matter.

2. We say that $I \triangleleft R$ is a *T.I. ideal (trivial intersection ideal)* if for all $x \in G$ either $I^x = I$ or $I^x \cap I = 0$. Such ideals come about as follows.

LEMMA 2.6. *Assume that R is G-semiprime and that $I \triangleleft R$. Then I is a T.I. ideal if and only if $\sum_{x \in G/G_I} I^x$ is a direct sum.*

Here we use G/G_I to denote a set of right coset representatives for G_I in G. Notice that if the above sum is direct, then G permutes the summands in the same way that G permutes the cosets of G_I. Our second claim is that $H = G_I$ for some T.I. ideal I. Note that $I^x \cap I = 0$ implies that $I^x I = 0$. Thus if R is a prime ring, then every T.I. ideal is G-stable and again this implies that the primeness or semiprimeness of $R*G$ depends on the finite normal subgroups of G. This explains an old result due to Montgomery and me.

The precise necessary and sufficient conditions for $R*G$ to be prime or semiprime are rather ugly. So we will just mention some corollaries which are analogous to the group algebra results.

THEOREM 2.7. *Suppose that $R*G$ is given with R a G-prime ring. If $\Delta^+(G_I) = 1$ for all T.I. ideals I, then $R*G$ is prime.*

THEOREM 2.8. *Let $R*G$ be given with R a G-semiprime ring. If $\Delta^p(G_I) = 1$ for all T.I. ideals I and all primes p such that I has p-torsion, then $R*G$ is semiprime.*

Let me also mention another result which is somewhat easier to prove. It has applications to the zero divisor problem.

PROPOSITION 2.9. *Let $R*G$ be given with R a domain. If $R*N$ is a domain for all finite normal subgroups N of G, then $R*G$ is prime.*

The first theorem above finesses the finite group problem by assuming that all such groups involved have order 1. The second result uses a version of Maschke's Theorem. Let us briefly consider what happens when G is finite.

If $|G| < \infty$, then the group algebra $K[G]$ is never prime unless $G = 1$. On the other hand, for twisted group algebras $K^t[G]$, the analogous question is decidedly nontrivial and has been studied by group theorists under the name *groups of central type*. Using the classification of all finite simple groups, Howlett and Isaacs have recently shown that if $K^t[G]$ is prime and if K is algebraically closed of characteristic 0, then G must be a solvable group. It is also known that the primeness of $K^t[G]$ reduces to questions about Sylow subgroups. Lorenz and I have an analogous reduction for crossed products $R*G$.

By Maschke's Theorem, $K[G]$ is semiprime if and only if $|G| \neq 0$ in K. For crossed products, there is the lovely analog of Fisher and Montgomery which asserts that if R is semiprime with no $|G|$-torsion, then $R*G$ is semiprime. Note that, if $R*G$ is semiprime, then it can be shown that $R*H$ is also semiprime for all subgroups H of G. Conversely if $R*H$ is semiprime for $H = 1$ and for all elementary abelian p-subgroups of G for which R has p-torsion, then $R*G$ is semiprime.

Now let us consider G-graded rings. To start with, the crossed product techniques work just fine for strongly graded rings and yield analogous results. But to go further requires a new trick – *duality*. Suppose first that G is finite. If S is a G-graded ring, then there is a ring $S\#G$ generated by S and certain orthogonal idempotents indexed by the elements of G. This is a fairly easy extension to deal with and it turns out that G acts as automorphisms on this ring. Thus we can form the skew group ring $(S\#G)G$.

THEOREM 2.10. (Cohen-Montgomery). *If G is a finite group of order n and if S is a G-graded ring, then $(S\#G)G \cong \mathrm{M}_n(S)$.*

I like Beattie's notation which uses $G*S$ for $S\#G$. With this, the above duality looks like $(G*S)*G \cong \mathrm{M}_n(S)$. As a consequence, we have

COROLLARY 2.11. (Cohen-Montgomery). *Let S be a G-graded ring with no $|G|$-torsion. If S is graded semiprime, then it is a semiprime ring.*

Notice that, in this result, assumptions on the graded structure of S replace the usual assumptions on the identity component S_1. These ideas were then extended to infinite groups by Quinn, who also determined how arbitrary subgroups fit into the duality picture. As a consequence, he obtained necessary and sufficient conditions for suitable group-graded rings to be prime or semiprime in the infinite case. These results are unfortunately somewhat messy to state.

We close this section with one last mention of finite groups. As we said earlier, Maschke's Theorem really concerns modules and the following crossed product formulation is due to D. G. Higman. Suppose $W \subseteq V$ are $R*G$-modules and assume that $W \mid V$ as R-modules. If multiplication by $|G|$ is a bijection on V, then $W \mid V$ as $R*G$-modules. There is also an *essential version* of this result. Again let $W \subseteq V$ be $R*G$-modules and assume that W is essential as an $R*G$-submodule of V. If V has no $|G|$-torsion, then W is essential as an R-submodule of V. This can be used to give an alternate proof of the Fisher-Montgomery Theorem. Finally, there is a *graded version* due independently to Quinn and Năstăsescu. Here we suppose that S is a G-graded ring and that $W \subseteq V$ are S-modules with no $|G|$-torsion. If W is essential in V as an S-module and if $0 \neq v \in V$, then $vS_y \cap W \neq 0$ for some $y \in G$. In particular, if $\ell_V(S_x) = 0$ for all $x \in G$, then W is essential in V as an S_1-module.

§3. Prime ideals

We now consider prime ideals in crossed products and, as is to be expected, the results here are much less general. If P is a prime ideal of $R*G$, then $I = P \cap R$ is a G-*prime ideal* of R. By this we mean that I is a G-stable ideal of R and that R/I is a G-prime ring. Notice that $I*G$ is a well understood ideal of $R*G$ and that $(R*G)/(I*G) \cong (R/I)*G$. Furthermore, $\bar{P} = P/(I*G)$ a prime ideal in this new crossed product with $\bar{P} \cap (R/I) = 0$. Thus without loss of generality we can study prime ideals P of $R*G$ with $P \cap R = 0$ and in particular we can now assume that R is a G-prime ring.

We will further assume that either (f) G is finite (joint work with Lorenz), or (p) G is polycyclic-by-finite and R is a right Noetherian ring. Here *polycyclic-by-finite* means that G has a finite subnormal series

$$1 = G_0 \triangleleft G_1 \triangleleft \cdots \triangleleft G_n = G$$

with each quotient G_{i+1}/G_i either infinite cyclic or finite. A version of the Hilbert Basis Theorem then implies that the case (p) crossed products are all right Noetherian.

Under either assumption, if Q is a minimal prime of R, then Q has only finitely many distinct G-conjugates. In particular, if $H = G_Q$ then H is a subgroup of G of finite index. Furthermore, since R is G-prime it follows that $\bigcap_{g \in G} Q^g = 0$.

THEOREM 3.1. *With the above assumptions and notation, there exists a precisely defined one-to-one-to-one correspondence between*

(i) *the prime ideals P of $R*G$ with $P \cap R = 0$,*

  (ii) *the prime ideals T of $R*H$ with $T \cap R = Q$, and*
  (iii) *the prime ideals $\bar{T}$ of $(R/Q)*H$ with $\bar{T} \cap (R/Q) = 0$.*

Since R/Q is a prime ring, this result effectively reduces the problem to the case of prime coefficient rings and at this point localization comes into play.

Let R be a prime ring and let S denote its *symmetric Martindale ring of quotients*. Then S satisfies

  (a) $S \supseteq R$ with the same 1.
  (b) If $q \in S$, then there exist $0 \neq A, B \lhd R$ with $Aq, qB \subseteq R$.
  (c) If $0 \neq q \in S$ and $0 \neq I \lhd R$, then $Iq \neq 0$ and $qI \neq 0$.
  (d) Let $0 \neq A, B \lhd R$ and let $f: {}_R A \to {}_R R$ and $g: B_R \to R_R$ be appropriate R-module homomorphisms. If f and g satisfy the *balanced* or *associative* condition $(af)b = a(gb)$ for all $a \in A$ and $b \in B$, then there exists $q \in S$ with $af = aq$ and $gb = qb$ for all $a \in A, b \in B$.

It turns out that such a ring extension S exists and that it is uniquely characterized by these properties. Furthermore, S is a prime ring and $C = \mathbb{Z}(S)$ is a field called the *extended centroid* of R. Note that properties (b) and (c) assert that S is somehow close to R, while condition (d) forces S to be reasonably large.

Now suppose $R*G$ is any crossed product with R a prime ring. Then, following the work of Fisher and Montgomery, there are a number of observations which can be made. First, the action of each element of G on R extends uniquely to an automorphism of S and therefore there exists a uniquely determined crossed product $S*G$ containing $R*G$. Next, we define $G_{\mathrm{inn}} = \{ g \in G \mid \bar{g} \text{ is inner on } S \}$. Thus G_{inn} is the set of elements of G which induce *X-inner automorphisms* on R. Furthermore, G_{inn} is a normal subgroup of G and $S*G_{\mathrm{inn}} = S \otimes_C C^t[G_{\mathrm{inn}}]$ where $C^t[G_{\mathrm{inn}}] = \mathbb{C}_{S*G}(S)$ is some twisted group algebra of G_{inn} over the extended centroid C. In addition, G acts as automorphisms on $C^t[G_{\mathrm{inn}}]$.

THEOREM 3.2. *Let $R*G$ satisfy either (f) or (p) and assume that R is a prime ring. Then, with the above notation, there is a precisely defined one-to-one-to-one correspondence between*

  (i)   *the prime ideals P of $R*G$ with $P \cap R = 0$,*
  (ii) *the prime ideals P' of $S*G$ with $P' \cap S = 0$, and*
  (iii) *the G-prime ideals T of $C^t[G_{\mathrm{inn}}]$.*

Combining the latter two results, we see that the prime ideals of $R*G$ depend on the H-prime ideals of a certain twisted group algebra $C^t[N]$.

Furthermore, if G is finite, then $C^t[N]$ is a finite dimensional algebra and if G is polycyclic-by-finite, then N is also polycyclic-by-finite and $C^t[N]$ is both right and left Noetherian. In either case, it follows that any H-prime ideal T of $C^t[N]$ is a finite intersection of the form $T = \bigcap_{g \in G} Q^g$ where Q is a prime ideal of $C^t[N]$ minimal over T. Thus we see that the prime ideals of $R*G$ eventually depend on the prime ideals of a certain twisted group algebra.

Our next step, therefore, is to consider the prime ideals in twisted group algebras $K^t[G]$ with G either finite or polycyclic-by-finite. In the former case, $K^t[G]$ is a finite dimensional algebra, so we will assume that its prime ideals are known. Thus only the polycyclic-by-finite groups are of interest and we start by describing the wonderful work of Roseblade in the case of ordinary group algebras.

1. If I is any ideal of $K[G]$, we define $I^\dagger = \{ g \in G \mid 1 - g \in I \}$. Notice that G is a subset of $K[G]$ and that under the homomorphism $K[G] \to K[G]/I$ we have $G \to G/I^\dagger$. Thus $I^\dagger$ is a normal subgroup of G. Furthermore, we say that I is *faithful* if $I^\dagger = 1$ and that I is *almost faithful* if $I^\dagger$ is finite. Since the ring epimorphism $K[G] \to K[G/I^\dagger]$ is well understood and since I is the complete inverse image of a faithful ideal in the latter group algebra, it usually suffices to study faithful ideals. So let us describe the faithful primes.

2. Recall that Zalesskiĭ's Theorem (Theorem 1.4) asserts that a particular prime ideal P of $K[G]$ is controlled by $\mathbb{Z}(G)$. Here G is a torsion free nilpotent group and such groups satisfy $\mathbb{Z}(G) = \Delta(G)$. Thus we might conjecture that, for G polycyclic-by-finite, all faithful primes of $K[G]$ are controlled by $\Delta(G)$. Unfortunately this is false as stated, but it is close to being true. It turns out that G has a characteristic subgroup G_0 of finite index such that all faithful prime ideals of $K[G_0]$ are controlled by $\Delta(G_0)$. The definition of G_0 is important, but somewhat complicated. First, a subgroup H of G is *orbital* if H has only finitely many G-conjugates or equivalently if $|G : \mathsf{N}_G(H)| < \infty$. Next, H is an *isolated orbital subgroup* if H is orbital and if any larger orbital subgroup L satisfies $|L : H| = \infty$. Finally $G_0 = \mathrm{nio}(G)$ is the intersection of the normalizers of all the isolated orbital subgroups of G.

We can combine the above two ideas as follows. We say that P is a *standard prime ideal* of $K[G]$ if $P = (P \cap K[\Delta])K[G]$ and $P \cap K[\Delta] = \bigcap_{g \in G} Q^g$ for some almost faithful prime ideal Q of $K[\Delta]$. It is easy to see that any standard prime is necessarily almost faithful. More generally, we say that P is *virtually standard* if the image of P in $K[G/P^\dagger]$ is a standard prime.

THEOREM 3.3. (Roseblade). *Let G be a polycyclic-by-finite group and let K be any field. Then $G_0 = \mathrm{nio}(G)$ is a characteristic subgroup of G of finite index and all prime ideals of $K[G_0]$ are virtually standard.*

If G is polycyclic-by-finite, then it is easy to see that $\Delta(G)$ is both finite-by-abelian and center-by-finite. In other words, $\Delta(G)$ is almost abelian. As a consequence of this reduction and known results for polynomial rings, Roseblade was then able to compute the prime and primitive lengths of $K[G]$ and he could determine which prime ideals were primitive. Futhermore, he proved that if $G = \mathrm{nio}(G)$, then $K[G]$ is a catenary ring.

Lorenz and I wanted to show that $K[G]$ is catenary for all polycyclic-by-finite groups G and the previous result gets us quite close. Indeed, if $G_0 = \mathrm{nio}(G)$, then $K[G] = K[G_0]*(G/G_0)$, G/G_0 is finite and the structure of the prime ideals of $K[G_0]$ is known. Thus we were motivated to study the prime ideals in crossed products $R*W$ with W finite and we hoped to obtain a result which would lift the information we require. As it turned out, this result enabled us to describe the prime ideals of $K[G]$ in general. I won't give the complete description here, but rather I will just indicate some of the ingredients.

First, any prime ideal P of $K[G]$ has a *vertex* N which is an isolated orbital subgroup of G. In particular, the normalizer of N has finite index in G and N is uniquely determined by P up to G-conjugation. Next, P has a *source* S which is a rather special type of ideal of the group algebra $K[\nabla_G(N)]$, where $\nabla_G(N)/N = \Delta(\mathbb{N}_G(N)/N)$. Again, this source S is appropriately unique and we have $P = S^G$ where, in this context, S^G denotes the largest two-sided ideal of $K[G]$ contained in $S \cdot K[G]$. In this way, we obtain a precise description of the prime ideals of $K[G]$. Unfortunately this did not lead to a solution of the catenary problem and in fact this problem is still open. We remark that Kaplansky has an example of a commutative crossed product $R*G$ with $|G| = 2$ such that R is a catenary ring but $R*G$ is not.

It remains to consider the prime ideals in twisted group algebras $K^t[G]$ and here we use the fact that polycyclic-by-finite groups are finitely presented. It then follows easily that there exists a polycyclic-by-finite group $\tilde{G}$ and a natural epimorphism $K[\tilde{G}] \to K^t[G]$. Furthermore $\tilde{G}/\tilde{Z} \cong G$ where $\tilde{Z}$ is a central subgroup of $\tilde{G}$ and the epimorphism maps $K[\tilde{Z}]$ to K. In particular, any prime ideal P of $K^t[G]$ lifts to a prime $\tilde{P}$ of $K[\tilde{G}]$. But then $\tilde{P}$ can be described via the mechanism of the preceding paragraph and this yields an appropriate description for P. There is however one slight difficulty with this idea, namely $\tilde{G}$ is not uniquely determined by the twisted group algebra. Thus, we do not know apriori which subgroup of G corresponds to $\mathrm{nio}(\tilde{G})$. Fortunately, there is a way around this.

PROPOSITION 3.4. *Let G be a polycylic-by-finite group and set $\mathrm{nio}^2(G) = \langle x^2 \mid x \in \mathrm{nio}(G) \rangle$. Then $\mathrm{nio}^2(G)$ is a characteristic subgroup of G of finite index. Furthermore, suppose $\tilde{G}$ is any polycyclic-by-finite group with $\tilde{G}/\tilde{Z} = G$ and $\tilde{Z} \subseteq \mathbb{Z}(\tilde{G})$. If $\tilde{H}/\tilde{Z} = \mathrm{nio}^2(G)$, then we have $\tilde{H} = \mathrm{nio}(\tilde{H})$ and hence $\tilde{H} \subseteq \mathrm{nio}(\tilde{G})$.*

With this result in place, it is clear that we can now unambiguously describe the prime ideals of $K^t[G]$. The description is of course quite complicated, so we will not pursue this line of reasoning any further.

§4. Krull relations

As a consequence of the characterization of prime ideals in crossed products of finite groups, one obtains the following *Krull relations*. This is again joint work with Lorenz.

THEOREM 4.1. *Let $R*G$ be given with G a finite group.*

(i) *(Cutting Down) If $P \lhd R*G$, then $P \cap R = \bigcap_{x \in G} Q^x$ where $Q \lhd R$ is unique up to G-conjugation. We say that P lies over Q.*

(ii) *(Lying Over) If $Q \lhd R$, then there exist $P_1, P_2, \ldots, P_n \lhd R*G$, with $1 \leq n \leq |G|$, such that each P_i lies over Q.*

(iii) *(Incomparability) If $P \subset P'$ are prime ideals of $R*G$, then we have $P \cap R \subset P' \cap R$.*

(iv) *(Going Up and Down) All four versions are satisfied.*

As we now know, this result actually holds in a far more general context. To be precise, we say that $R \subseteq S$ is a *finite normalizing extension* if R and S are rings with $S = \sum_{i=1}^{m} Rx_i$ and $Rx_i = x_i R$ for all i. Thus, for example, if G is finite, then $R \subseteq R*G$ is a finite normalizing extension since $R*G = \sum_{g \in G} R\bar{g}$ and $R\bar{g} = \bar{g}R$ for all $g \in G$.

THEOREM 4.2. (Heinicke-Robson). *Let $S = \sum_{i=1}^{m} Rx_i$ be a finite normalizing extension of R.*

(i) *(Cutting Down) If $P \lhd S$, then $P \cap R = \bigcap_{i=1}^{n'} Q_i$ is a finite intersection of $n' \leq m$ minimal covering primes. We say that P lies over each Q_i.*

(ii) *(Lying Over)* If $Q \lhd R$, then there exist $P_1, P_2, \ldots, P_n \lhd' S$, with $1 \leq n \leq m$, such that each P_i lies over Q.

(iii) *(Incomparability)* If $P \subset P'$ are prime ideals of S, then $P \cap R \subset P' \cap R$.

(iv) *(Going Up and Down)* All four versions hold.

Heinicke and Robson also consider intermediate rings. Here $R \subseteq T \subseteq S$ with $R \subseteq S$ a finite normalizing extension. They study both *intermediate extensions* $R \subseteq T$ and $T \subseteq S$ and they obtain various analogs of the previous theorem.

Finally, Cohen and Montgomery use duality to obtain similar results for group-graded rings. Here S is a G-graded ring with G finite and their results relate the primes of S to the prime ideals of the identity component S_1 and to the graded prime ideals of S. One version of Going Up, however, fails in this context.

In the waning moments of this talk, let me briefly mention one application of all of this to the Galois theory of rings. Thus suppose G is a finite group which acts on a ring R and assume that $|G|^{-1} \in R$. Then we can form the skew group ring RG and we let $e = |G|^{-1} \sum_{g \in G} g \in RG$. It follows easily that e is an idempotent of RG with $eRGe = R^G e \cong R^G$, where R^G is the fixed ring. Furthermore, we note the classical fact that, for any ring S and idempotent $f \in S$, there exists a one-to-one correspondence between the prime ideals of fSf and the primes of S not containing f. Thus, combining all these ingredients yields

THEOREM 4.3. (Montgomery). *Assume that G acts on R with $|G|^{-1} \in R$.*

(i) *(Cutting Down)* If $P \lhd R$, then $P \cap R^G = \bigcap_{i=1}^{n} Q_i$ is a finite intersection of $n \leq |G|$ minimal covering primes. We say that P lies over each Q_i.

(ii) *(Lying Over)* If $Q \lhd' R^G$, then there exists $P \lhd R$, unique up to G-conjugation, such that P lies over Q.

(iii) *(Incomparability)* If $P \subset P'$ are prime ideals of R, then we have $P \cap R^G \subset P' \cap R^G$.

(iv) *(Going Up and Down)* Three of the four versions hold.

This completes our discussion of prime ideals and I would like to close with a few thoughts in other directions. Recall that this talk began with the epimorphism $\xi: K[G] \to R$ and the assumption that R has some reasonable ring theoretic structure. In particular, if R is prime, then the kernel of ξ is a prime ideal. But there are certainly other important ring theoretic properties and these give rise to other interesting questions. For example

(1) When is $K[G]$ a domain?

(2) When is $K[G]$ semiprimitive?

(3) What does the Jacobson radical of $K[G]$ look like?

(4) When does $K[G]$ satisfy a polynomial identity?

Question (4) was settled quite a while ago and Lorenz will discuss (1) in his talk. Zalesskiĭ and I spent about ten years on problems (2) and (3) and we had at least some success. These two problems are now unfortunately dormant and I would surely love to see them resurrected. But that's another story and the theme for quite a different sort of talk.

SELECTED REFERENCES

1. R. B. Howlett and I. M. Isaacs, *On groups of central type*, Math. Z. **179** (1982), 555–569.
2. J. C. McConnell and J. C. Robson, *"Noncommutative Noetherian Rings,"* Wiley-Interscience, New York, 1987.
3. S. Montgomery, *"Fixed Rings of Finite Automorphism Groups of Associative Rings,"* Lecture Notes in Math. 818, Springer, Berlin, 1980.
4. D. S. Passman, *"The Algebraic Structure of Group Rings,"* Wiley-Interscience, New York, 1977, (Krieger, Malabar, 1985).
5. D. S. Passman, *Group rings of polycyclic groups*, in *"Group Theory: essays for Philip Hall,"* Academic Press, London, 1984, pp. 207–256.
6. D. S. Passman, *"Infinite Crossed Products,"* Academic Press, Boston, 1989.
7. J. E. Roseblade, *Prime ideals in group rings of polycyclic groups*, Proc. London Math. Soc. (3) **36** (1978), 385–447.

Department of Mathematics
University of Wisconsin-Madison
Madison, Wisconsin 53706

Category Theory and Quantum Field Theory

Abstract. We describe the recent basic result of Doplicher and Roberts characterizing the category of representations of a compact group. We indicate their motivation, coming from quantum field theory, and then state some important related category-theoretic questions which have arisen recently in low-dimensional quantum field theory. These questions, which are quite open, have interesting connections with quantum groups and knot theory.

§1. Introduction

I'm sure you can believe that I enjoyed choosing this somewhat improbable title. But in fact there are at present leading theoretical physicists [**MS**] who are devoting sections of their papers to aspects of category theory, raising compelling questions and proving fine theorems, of a kind which I believe will be of some interest to non-commutative ring theorists. In my two lectures I would like to describe some of these developments to you. In the first lecture I will describe a fundamental question which emerged about twenty years ago, but which received a satisfactory answer only very recently. Then in my second lecture I will describe some questions which have come to the forefront only quite recently, and remain quite unanswered. We will see that not only are some quite interesting non-commutative rings involved, but there are fascinating connections with quantum groups, braid groups, and invariants of knots and links.

First, however, I must tell you that I have never conducted research in this subject (though it is not so far from my own research interests), but rather I have been a fascinated observer of these developments for about twenty years. Since I am not an expert in the subject, I must warn you that my knowledge of some parts of what I will tell you is quite superficial.

Also, the panorama which I will try to cover is quite vast, far too vast to be described precisely in two short lectures. Thus I will be oversimplifying substantially. In doing this, I will try to concentrate on those features which I believe to be of most interest to non-commutative ring theorists. Many of the results which I will discuss will not be stated precisely, since I will not have taken the time to explain all of the necessary hypotheses. But it is my hope that this over-simplification will result in the overall picture being easier for non-experts to grasp.

§2. The physical set-up

How do physicsts model a physical situation involving both quantum mechanics and special relativity? (No one knows how to do it with general relativity.) There are several fairly different looking, but essentially equivalent, formulations. (See [**GJ**] for precise statements and comparisons.) I will give a rough description of the formulation which is closest to ring-theory. This is commonly known as algebraic quantum field theory. For simplicity of exposition, I will describe the formulation only for physical systems in which just short-range forces are being considered, e.g. scattering problems, and in which there are no massless particles (like photons).

In algebraic quantum field theory, one models a given such system by means of a *-algebra (algebra with involution, over the complex numbers), A, whose elements are called the local bounded observables. To every bounded open region, $\mathcal{O}$, of flat space-time, $R^d \times R$, there is to be associated a *-subalgebra, $A(\mathcal{O})$, consisting of the observables which are measurable in $\mathcal{O}$. This association is to be inclusion-preserving, and A is to be the union of all of the $A(\mathcal{O})$'s. As a rough indication of where such a structure might come from, let $V = C_c^\infty(R^{d+1})$, and let A be, depending on the situation, the tensor algebra, or the symmetric tensor algebra, or the antisymmetric tensor algebra, over the vector space V; then let $A(\mathcal{O})$ be the subalgebra corresponding to the subspace $C_c^\infty(\mathcal{O})$.

Now on R^{d+1} we have the Lorentz metric, say with d plus signs and one minus sign. (We will see that it is of much interest to consider what happens as the dimension, d, of space varies.) Then two points of space-time are said to be *space-like separated* if their inner-product for this metric is strictly positive. Two regions will then be said to be *space-like separated* if each point in one is space-like separated from each point in the other. The crucial axiom which gives the theory its flavor is Einstein Causality, which asserts that measurements made in two space-like separated regions can not influence each other, so that two observables which are measurable in two space-like separated regions are simultaneously observable. But this simultaneous observability is encoded exactly by having the corresponding elements of A commute. Thus Einstein Causality says that if $\mathcal{O}_1$ and $\mathcal{O}_2$ are space-like separated regions, then all elements of $A(\mathcal{O}_1)$ commute with all elements of $A(\mathcal{O}_2)$.

For a given region $\mathcal{O}$, its space-like complement, $\mathcal{O}'$, consists, by definition, of all the points which are space-like separated from all the points in $\mathcal{O}$. It is useful to define $A(\mathcal{O}')$ to be the subalgebra generated by the union of all the $A(\mathcal{O}_1)$'s for which $\mathcal{O}_1$ is space-like separated from $\mathcal{O}$. Again, these two subalgebras are in each other's commutants.

There are a number of additional axioms. The generally accepted list is known as the Haag-Kastler axiom system, since Haag and Kastler [**HK**] in 1964 first pulled them together as the basic ones. They include, for example, the principle that the Poincaré group should act as a group of automorphisms of A satisfying certain properties. But for our purposes it will be sufficient to require that there be an action, α, of the translation group R^{d+1} on A which is consistent with its action on space-time, that is, such that for any $x \in R^{d+1}$ and any region $\mathcal{O}$, we have $\alpha_x(A(\mathcal{O})) = A(\mathcal{O} + x)$.

As is usual in quantum physics, the states of the system being modeled will come from unit vectors in the Hilbert spaces of *-representations of the algebra A. A key distinction between the quantum mechanics of a finite number of particles on the one hand, and quantum field theory on the other, is that for the latter a single representation will not suffice. However, in the local theories which we are considering, there is to exist a distinguished *-representation, faithful and irreducible, called the *vacuum representation*, denoted by π_0. The only other physically relevant representations are then those which are local perturbations of π_0, that is, representations π such that there is a bounded region $\mathcal{O}$ such that the restrictions of π and π_0 to $A(\mathcal{O}')$ are unitarily equivalent. They are also required to be covariant for the action α, in the sense that there is a unitary representation, U, of R^{d+1} on the Hilbert space of π such that for any $a \in A$ and $x \in R^{d+1}$ we have

$$\pi(\alpha_x(a)) = U_x \pi(a) U_x^{-1}.$$

We will refer to the representations which satisfy these two conditions as the *physical representations*. Unitary equivalence classes of irreducible physical representations are called superselection sectors; for any model their classification is a basic concern.

Even for free systems (i.e. where there are no interactions between particles), it takes substantial work to construct actual examples of such structures [**GJ**]; while when interactions are present, it requires a major effort [**GJ**], and the situation is still not satisfactorily understood for space-time of dimension 4.

The starting point for the Doplicher-Haag-Roberts program was their observation [**DHR1**] that what physicists were actually doing when constructing specific models was not just directly constructing the algebra A of observables, but rather constructing a larger algebra, F, of "fields", which did not satisfy Einstein Causality, and so were recognized as not having direct physical meaning. But the physicists also had present a compact Lie group, G, such as $SU(3)$, called the *gauge group* (of the first kind), together with an action of G on F such that A is the fixed-point algebra for this action. A fundamental question then, which Doplicher, Haag, and Roberts set out to answer was:

QUESTION: Do G and F contain significant information beyond that contained in A, or are they, in fact, just useful auxiliary constructs which can be recovered from A itself?

For the case in which there are additional hypotheses which, in effect, force the gauge group to be Abelian, they succeeded in 1969 [**DHR2**] in showing that G and F can indeed be recovered from A. It is only very recently [**DR4**] that Doplicher and Roberts made the major breakthroughs which settle the situation for possibly non-Abelian gauge groups. But we will see shortly that there is a very interesting story beyond that.

§3. Endomorphisms

For our purposes the precise description of the physical representations is not needed, because one immediately transforms them into a structure which is more easily managed. Specifically, one uses the unitary operator from the Hilbert space for π to the Hilbert space for π_0 which gives the equivalence of the restrictions of these representations to some $A(\mathcal{O}')$, to identify these two Hilbert spaces. One then finds [**DHR3**] from the axioms, that π is of the form $\pi_0 \circ \rho$, where ρ is an endomorphism of A (which preserves the identity element). The endomorphisms thus obtained — the *physical endomorphisms* — have properties corresponding to those of the representations. These include locality in the sense that there is a bounded region $\mathcal{O}$ such that the restriction of ρ to $A(\mathcal{O}')$ is the identity endomorphism, and equivalence to their translates by α in the sense that for any $x \in R^{d+1}$ there is a unitary element, V_x, of A such that

$$\alpha_x \circ \rho \circ \alpha_{x^{-1}} = ad_{V_x} \circ \rho.$$

Even more, if $\mathcal{O}$ is a region in which ρ is localized as above, then V_x will be in the subalgebra corresponding to the region which is the union of all the translates of $\mathcal{O}$ by the points of any fixed path from 0 to x. This homotopy property is crucial for obtaining the further properties of the situation. Another crucial property of the physical endomorphisms which comes from the axioms, is that the composition of any two of them is again a physical endomorphism [**DHR3**]. Thus the physical endomorphisms form a semigroup with identity element.

The result of the above considerations is to focus attention on the endomorphisms of an algebra. Let A be any algebra (or ring), and let S be any collection of endomorphisms of A (preserving the identity element). Given ρ_1 and ρ_2 in S, let

$$Hom(\rho_1, \rho_2) = \{a \in A : a\rho_1(b) = \rho_2(b)a \ \text{ for all } \ b \in A\}.$$

This is a linear subspace of A, and it is easily checked that with these subspaces as the sets of "morphisms", S forms an additive category, where the "composition" of morphisms is given just by the product in A. The considerations which come next suggest the following moral, which, to the best of my knowledge, has not been much considered by ring-theorists:

MORAL: For a given ring, study of a suitable category of its endomorphisms may yield useful information about the structure of the ring.

Suppose now that S is actually a semigroup under composition. Given $a \in Hom(\rho_1, \rho_2)$ and $b \in Hom(\rho_1', \rho_2')$, it is immediately checked that $a\rho_1'(b)$ is in $Hom(\rho_1'\rho_1, \rho_2'\rho_2)$. With this additional operation, S becomes a strict monoidal category [**ML**], in the sense of having an operation which looks somewhat like the inner tensor product of group representations. The term "strict" here refers to the fact that the associative law holds not just as an equivalence of objects in the category, but as an actual equality.

What Doplicher, Haag, and Roberts saw very early [**DHR1**] is that in those examples constructed by means of a gauge group and algebra of fields, the category S of physical endomorphisms is equivalent to the category of all finite-dimensional unitary representations of the gauge group, with composition of endomorphisms corresponding to tensor product of representations. (I will give some indication of the reason in the next section.) This clearly suggests a strategy for approaching the fundamental question stated above, namely of trying to show that S is equivalent to the category of representations of some compact group, and of recovering this group in the process. Thus one is lead to the following attractive and basic category-theoretic question:

QUESTION: How does one characterize those categories which arise as the categories of finite-dimensional unitary representations of a compact (or even finite) group?

When asked this question, one probably thinks instantly of the Tannaka-Krein duality theorem for compact groups. But in that theorem, the objects are assumed to be vector spaces and the morphisms are assumed to be linear maps, whereas in the situation from physics described above, the objects are endomorphisms of an algebra, and have no vector space structure, while the *Hom* spaces, though possessing a vector space structure, do not consist of linear maps. Thus the Tannaka-Krein theorem does not provide an answer.

Of course, in seeking an answer to the above question, it is natural to assume at least that the category has an operation analogous to forming inner tensor products of representations, that is, is a monoidal category. Now a key property of taking actual inner tensor products of representations is that if π_1 and π_2 are representations of a group, then $\pi_1 \otimes \pi_2$ is naturally equivalent to $\pi_2 \otimes \pi_1$, and, even more, that if ϵ is a natural transformation consisting of morphisms (operators) implementing this equivalence, then it can be chosen such that

$$\epsilon(\pi_1, \pi_2)\epsilon(\pi_2, \pi_1) = 1.$$

A general monoidal category satisfying the analogue of this property, as well as the obvious properties one wants with respect to the monoidal identity element, is said to be *symmetric*. But in seeking an answer to the above question one must assume more, since the category of finite-dimensional representations of a semigroup will form a symmetric monoidal category. Now for a group, every representation has a contragradient representation, and so it is natural to assume that one's symmetric monoidal category has such a structure — especially since in the case from physics coming from a gauge group one sees easily that this additional structure is present. There are several closely related ways of formulating this additional structure in the abstract case, in terms of a *-structure on the category satisfying relatively strong properties. In particular, one consequence which one wants from any formulation is the semi-simplicity of the algebras $Hom(\rho, \rho)$, so that they are just finite direct sums of full matrix algebras over the complex numbers, C. We will refer to the original papers [**DR3**] for a precise formulation, and just refer to a symmetric monoidal *category with conjugates* when such a structure is present.

Of course, if such a category is to be equivalent to the full category of finite dimensional-representations of a compact group, then it should be closed under taking subobjects, and under the formation of finite direct sums of objects. Furthermore, one must ensure that there is no excess multiplicity, which can be done in a natural way by requiring that if ι is the identity object for the monoidal structure, then

$$Hom(\iota, \iota) \cong C.$$

Let us say that the category is *normalized* if it satisfies this last property. It is only very recently [**DR3**] that Doplicher and Roberts found a proof of the fact that the various conditions described above are sufficient, that is, they proved:

THEOREM. *Let S be a normalized symmetric strictly monoidal category with conjugates closed under forming subobjects and finite direct sums. Then there is a compact group, G, unique up to isomorphism, such that S is equivalent to the category of finite-dimensional unitary representations of G.*

The proof of this theorem, though quite lengthy, is very interesting, with aspects which should be of interest to non-commutative ring theorists. In the next section I will try to give an indication of some of the ideas involved. Then in the following section we will go back to the original question from physics and see that, though the above theorem and its proof provide major progress towards an answer, much of great interest remains to be understood.

Let me also remark that the above theorem should not be the last word on its subject. It leaves quite open what happens in the important case when the strictness condition is dropped. Also, it only considers the situation over the field C — leaving quite open, in particular, what happens over fields of prime characteristics where semisimplicity can fail.

§4. Ideas of the proof

We will suppose that S is a normalized symmetric strictly monoidal category with conjugates, closed under taking subobjects and direct sums. To begin the proof, we will use the only linear structure which is present, namely that on the Hom spaces. Let ρ be any element of S. For any positive integer n we let ρ^n denote the nth power of ρ for the monoidal structure of S (analogue of the nth inner tensor power of a representation). Let I_ρ denote the identity element in $Hom(\rho, \rho)$. We will use the symbol $\otimes$ to denote the operation on morphisms which comes from the monoidal structure. Then for any integer k, positive or negative, we have, as soon as the positive integer n is such that $k + n$ is positive, the natural map

$$Hom(\rho^n, \rho^{k+n}) \longrightarrow Hom(\rho^{n+1}, \rho^{k+n+1})$$
$$T \longrightarrow T \otimes I_\rho.$$

It can be shown that these are injective maps between finite-dimensional vector spaces. For each fixed k we will let O_ρ^k denote the inductive limit of these maps as n goes to infinity. (The use of O here is traditional, and has nothing to do with the use of $\mathcal{O}$ earlier to denote regions, also traditional.) It is easily seen that composition of morphisms gives maps

$$O_\rho^k \times O_\rho^j \longrightarrow O_\rho^{k+j}.$$

Let

$$O_\rho = \bigoplus_k O_\rho^k.$$

It is easily seen that with the maps given just before, O_ρ becomes an associative Z-graded algebra. Furthermore, there is a canonical endomorphism of O_ρ, denoted $\hat{\rho}$, defined by

$$\hat{\rho}(T) = I_\rho \otimes T.$$

Thus O_ρ, as an algebra with an endomorphism, provides a micromodel, in some sense, of the kind of situation discussed in the previous sections. In fact:

THEOREM. *Let S_ρ be the full subcategory of S with objects $(\iota, \rho, \rho^2, \dots)$. Then (for suitable axioms) the map sending $T \in S_\rho$ to $T \in O_\rho$ is a faithful full embedding of S_ρ into the category $End(O_\rho)$.*

The algebras O_ρ form a quite interesting class of algebras. The prototypes, which are somewhat analogous to Clifford algebras, have not, as far as I know, been looked at much by non-commutative ring theorists. They are constructed as follows. Let H be a finite-dimensional Hilbert space of dimension n at least 2. Then O_n is defined to be the universal *-algebra generated by H as a subspace, subject to the relations:

(a) For any ϕ and ψ in H, we have

$$\phi^* \psi = \langle \phi, \psi \rangle 1.$$

(b) If $\{\phi_k\}$ is an orthonormal basis for H, then

$$\sum_k \phi_k \phi_k^* = 1.$$

(Note that if $\|\phi\| = 1$ then $\phi^*\phi = 1$ so that $\phi\phi^*$ is an idempotent. Also if ϕ is orthogonal to ψ, then the idempotents $\phi\phi^*$ and $\psi\psi^*$ are orthogonal. Thus the left-hand-side of (b) is always an idempotent. It is easily seen that (b) is independent of the choice of basis.) The algebras O_n are called Cuntz algebras — they were introduced by him [C] in 1977. (See [CK] for an interesting generalization.) An important feature of the Cuntz algebras is that they have a canonical endomorphism, σ, defined by

$$\sigma(T) = \sum_k \phi_k T \phi_k^*.$$

(This is independent of the choice of orthonormal basis.)

Suppose now that G is a compact group, and that ρ is a unitary representation of G on H. Then ρ gives an action of G on O_n. Let O_ρ denote the fixed-point subalgebra. It is invariant under σ. Then O_ρ, O_n, and the action ρ provide a kind of minimal model of the algebra of observables, algebra of fields, and action of the gauge group, considered earlier. I believe that there has been little investigation of the algebras O_ρ for various representations ρ, but that they may be quite interesting, and perhaps somewhat reminiscent of the algebras arising in classical invariant theory (except that these are non-commutative).

For the case in which ρ is a faithful self-conjugate representation of G, Doplicher and Roberts [DR2] show that, upon taking all "tensor" powers of the restriction of σ to O_ρ , and then taking subobjects, one obtains a strict monoidal category which is equivalent to the category of representations of G.

The main result of [DR2] is that if one has a situation like that of the above theorem, one can view it as an action of a potential group dual, and then one can apply a very interesting new kind of crossed product construction to obtain a larger algebra, F, such that the group of automorphisms of F leaving A pointwise fixed is a compact Lie group G. Furthermore, all of the endomorphisms of A coming from S are then given by G-invariant finite-dimensional Hilbert spaces in F as in the definition of σ above, with the category of representations of G on these Hilbert spaces being equivalent to S. The general case of a normalized symmetric monoidal category with conjugates can then be treated by suitably pasting together copies of this construction, yielding an algebra F acted upon by a compact group G which no longer need be Lie.

Since the discussion and full proof of all the above runs over a number of articles totaling several hundred pages, it is clear that what I have said above leaves out a vast number of details.

§5. Back to physics

To apply the above considerations to the quantum field theory situation described in Section 2, one must, of course, verify that the properties which we required of our monoidal category are satisfied. It turns out that for us the most interesting of the required properties is that of symmetry. Let us give a "proof" of this property, in part in order to give an indication of the type of arguments which are common in the subject.

LEMMA. *If ρ_1 and ρ_2 are physical endomorphisms which are localized in bounded regions $\mathcal{O}_1$ and $\mathcal{O}_2$ which are space-like separated, then ρ_1 and ρ_2 commute.*

PROOF: It suffices to show that they commute on each $A(\mathcal{O})$. So let $\mathcal{O}$ be fixed. Translate $\mathcal{O}_1$ and $\mathcal{O}_2$ continuously in opposite directions in such a way that they become space-like to $\mathcal{O}$, and that the regions they sweep out remain space-like separated to each other. The corresponding translates of the ρ_j's are then of the form $ad_{U_j} \circ \rho_j$, and both these translates are the identity on $A(\mathcal{O})$. Furthermore, by the condition on the swept out regions, and the homotopy property mentioned early in Section 3, we see that U_1 and U_2 commute. Thus, on $A(\mathcal{O})$,

$$\rho_1\rho_2 = ad_{U_1^*}ad_{U_2^*} = ad_{U_2^*}ad_{U_1^*} = \rho_2\rho_1.$$

$\square$

PROPOSITION. *For any two physical endomorphisms ρ_1 and ρ_2 , the products $\rho_1\rho_2$ and $\rho_2\rho_1$ are equivalent.*

PROOF: Let ρ_j be supported in the bounded region $\mathcal{O}_j$. Choose a translate of $\mathcal{O}_2$ which is space-like separated from $\mathcal{O}_1$. Let the corresponding translate of ρ_2 be $ad_U\rho_2$, which commutes with ρ_1 by the previous lemma. Then

$$\rho_2\rho_1 = ad_{U^*}ad_U\rho_2\rho_1 = ad_{U^*}\rho_1 ad_U\rho_2 = ad_{U^*\rho_1(U)}\rho_1\rho_2.$$

This says exactly that $U^*\rho_1(U)$ is a unitary intertwining operator between $\rho_1\rho_2$ and $\rho_2\rho_1$. $\square$

Further argument of this kind shows that the intertwining operator found above does not change as the translate of $\mathcal{O}$ moves continuously (remaining space-like separated). This brings us to a crucial, but quite subtle, aspect of the theory. In space-time of dimension at least 3, the complement of the full (forward and backwards) solid light-cone is connected, while in dimension 2 it is not. A bit of arguing as above shows then that for dimension at least 3, the intertwining operator is unique. (Notice, in particular, that a choice of phase in U in the above expression cancels out.) This intertwining operator is called the "statistics operator", and is traditionally denoted by $\epsilon(\rho_1, \rho_2)$, or by just ϵ_ρ when the two arguments are equal. Then ϵ is a natural transformation with, for dimension at least 3, the property that $\epsilon_\rho{}^2 = 1$, just as happens for the category of representations of a compact group. That is, we have found that the monoidal category of physical endomorphisms is symmetric. The above arguments for this crucial fact were first given by Doplicher, Haag, and Roberts in [**DHR3**, Lemma 2.6]. It follows that the more recent results of Doplicher and Roberts which we have described in the previous sections apply, and so they are able to prove their fundamental result ([**DR4**]):

THEOREM. *Let A be an algebra of local observables as defined earlier. If the dimension of space-time is at least 3, then there is a unique gauge group G and an algebra of fields F on which G acts, such that A is the fixed point algebra, and the physical endomorphisms of A are related to the finite-dimensional unitary representations of G in the manner indicated near the beginning of this article.*

§6. Question posed

But what happens if the dimension of space-time is 2? In this case examples show [**Fro**], [**S1**] that it may be impossible to arrange that $(\epsilon_\rho)^2$ be 1. Let us call a monoidal category with a natural transformation ϵ as above, but for which this equality fails, "quasi-symmetric" (so forming a "quasitensor" category in the terminology of [**Ma1**]). We saw that Doplicher and Roberts showed, in the sense described above, that when symmetry holds, together with some other natural conditions, one has the representation theory of a compact group. Accordingly, we see that a fundamental problem is to extend the theory of Doplicher and Roberts to obtain an answer to:

BASIC QUESTION: How does one describe the mathematical objects whose representation theory is a monoidal category with conjugates which satisfies quasi-symmetry.

In fact, it appears to me that a substantial amount of the recent work in conformal quantum field theory (e.g. [**FFK**], [**MS**]) can be interpreted as attempts to find the corresponding objects in the case of specific examples which have been constructed (with the name "chiral algebras" often used for these objects).

Some possibilities are known. There are the compact quantum groups [**W1–4**], the quantum enveloping algebras [**Dr**], [**Ji**], and other Hopf algebras. It would be important to obtain a good understanding of exactly which of these have quasi-symmetric representation theories [**Ma1–2**]. However, there is evidence [**S1–3**] that quantum groups and enveloping algebras will not suffice. But there are also the paragroups of Ocneanu [**O**]; these are not well understood except, to some extent, the ones which are analogous just to finite groups. There are also the hypergroups of Sunder [**Su**]. But it is not at all clear whether one can even expect all of these objects to be sufficient. (And, of course, one also has the interesting question of describing the objects whose representation theories are monoidal, but do not even satisfy quasi- symmetry (though they may satisfy various other conditions). Or of characterizing those monoidal categories whose representation theory is equivalent to that of a Hopf algebra, for instance. (For an extension of the classical Tannaka-Krein duality theorem to Hopf algebras, see [**U**].) It seems likely that in the process of answering these questions, much interesting algebra will be developed, such as, for example, possible construction of some analogues of the algebras O_ρ considered earlier. There is also evidence that categories of bimodules are likely to play a very useful role in these matters [**L2**], [**O**], [**S3**], [**Su**].

These considerations have some fascinating and unexpected connections with other subjects, such as the Jones index and knot theory. Given a physical endomorphism ρ, it is natural to ask for the "dimension" of the representation to which it might correspond. Choose a bounded region $\mathcal{O}$ in which ρ is localized. From the axioms of the local quantum field theories which we have been considering, it follows that there is a conditional expectation of $A(\mathcal{O})$ onto its subalgebra $\rho(A(\mathcal{O}))$. But these are appropriate conditions for defining the Jones index, as Watatani has recently shown in considerable generality [**Wa**]. Very recently Longo has shown [**L1–2**] that if one sets $dim(\rho) = (Index(\rho(A(\mathcal{O})), A(\mathcal{O}))^{1/2}$, then this is independent of the choice of $\mathcal{O}$, and does give the dimension of the corresponding representation of the gauge group when this is present. So this should be the good definition of dimension in general. But the fact that the Jones index need not be an integer provides an interesting phenomenon which one will have to deal with in trying to answer the basic problem posed above.

Turning to knots, it is not difficult to show [**Fr**], [**FRS**] that in the quasi-symmetric case, the ϵ defined above must satisfy

$$\epsilon_\rho \rho(\epsilon_\rho)\epsilon_\rho = \rho(\epsilon_\rho)\epsilon_\rho\rho(\epsilon_\rho).$$

If we apply ρ^{j-1} to this equation, and then set $\sigma_j = \rho^{j-1}(\epsilon_\rho)$, we see that the σ_j's satisfy the well-known braid relation

$$\sigma_j \sigma_{j+1} \sigma_j = \sigma_{j+1} \sigma_j \sigma_{j+1}.$$

They also satisfy the braid relation

$$\sigma_j \sigma_k = \sigma_k \sigma_j, \qquad |j - k| > 1.$$

In the case in which symmetry holds, we also have the relation $\sigma_j^2 = 1$, so that the σ_j's give a representation of the permutation group. In this case, one frequently says that the category satisfies "permutation symmetry". But if only quasi-symmetry holds, then one only obtains a representation of the braid group, in which case one frequently says that the category satisfies "braid symmetry".

Now for any physical endomorphism ρ_1, note that

$$Hom(\rho_1, \rho_1) = (\rho_1(A))',$$

where $'$ denotes taking commutants in A. Thus $\sigma_j \in (\rho^{j+1}(A))'$. But it is clear that for any k

$$(\rho^k(A))' \subseteq (\rho^{k+1}(A))'.$$

Thus we can view $\sigma_1, \ldots, \sigma_j$ as giving a representation of the braid group B_j into the unitary operators in $(\rho^{j+1}(A))'$.

Let E denote the conditional expectation from A to $\rho(A)$ mentioned above. Then $\phi = \rho^{-1}E$ will be a linear left inverse for ρ, which will have nice algebraic properties if E has been properly chosen [**DHR4**], [**FRS**]. In particular, one finds then that

$$\phi^j((\rho^j(A))' \subseteq \Lambda'.$$

But $A' = C$ in the cases of interest, since $A' = Hom(\iota, \iota)$, so that this is just the normalization condition used in Section 4. If we then set $\tau(b) = \phi^j(b)$ for $b \in (\rho^j(A))'$, then it can be shown [**FRS**] that this definition is coherent for the inclusions of $(\rho^j(A))'$ in $(\rho^{j+1}(A))'$ (which are finite-dimensional), and that τ then defines a trace on $\bigcup(\rho^j(A))'$ which has the Markov property that insures that when τ is restricted to (the image of) the braid groups, it gives a link invariant (after suitable normalizations).

In this way, one arrives at the astonishing fact that every 2-dimensional local quantum field theory gives an invariant of knots and links. Notice that the axiom primarily responsible for this is Einstein Causality. One of the first link invariants which one can obtain in this way is the Jones polynomial [**Jo**], and there is presently intensive study of the further link invariants which arise from various specific models of quantum field theories. (There are also some very interesting non-local models in space-time of dimension 3 for which similar considerations are possible [**FGM1–2**], [**FK**].)

REFERENCES

[C] J. Cuntz, *Simple C*-algebras generated by isometries*, Comm. Math. Phys. **57** (1977), 173–185.

[CK] J. Cuntz and W. Krieger, *A class of C*-algebras and topological Markov chains*, Invent. Math. **56** (1980), 251–268.

[DHR1] S. Doplicher, R. Haag and J. E. Roberts, *Fields, observables and gauge transformations I*, Comm. Math. Phys. **13** (1969), 1–23.

[DHR2] S. Doplicher, R. Haag and J. E. Roberts, *Fields, observables and gauge transformations II*, Comm. Math. Phys. **15** (1969), 173–200.

[DHR3] S. Doplicher, R. Haag and J. E. Roberts, *Local observables and particle statistics I*, Comm. Math. Phys. **23** (1971), 199–230.

[DHR4] S. Doplicher, R. Haag and J. E. Roberts, *Local observables and particle statistics II*, Comm. Math. Phys. **35** (1974), 49–85.

[DR1] S. Doplicher and J. E. Roberts, *Duals of compact Lie groups realized in the Cuntz algebras and their actions on C*-algebras*, J. Funct. Anal. **74** (1987), 96–120.

[DR2] S. Doplicher and J. E. Roberts, *Endomorphisms of C*-algebras, cross products and duality for compact groups*, Ann. Math. **130** (1989), 75–119.

[DR3] S. Doplicher and J. E. Roberts, *A new duality theory for compact groups*, Invent. Math. **98** (1989), 157–218.

[DR4] S. Doplicher and J. E. Roberts, *Why there is a field algebra with a compact gauge group describing the superselection structure in particle physics*, Comm. Math. Phys. (to appear).

[Dr] V. G. Drinfeld, *Quantum groups*, in "Proceedings of the International Congress of Mathematicians, Berkeley," Amer. Math. Soc., Providence, R. I., 1987, pp. 798–820.

[FFK] G. Felder, J. Fröhlich, and G. Keller, *On the structure of unitary conformal field theory II: Representation theoretic approach*, Comm. Math. Phys. (to appear).

[Fr] K. Fredenhagen, *Structure of superselection sectors in low-dimensional quantum field theory*, in "Proceedings of the XVII International Conference on Differential Geometric Methods in Theoretical Physics: Physics and Geometry, 1989" (to appear).

[FRS] K. Fredenhagen, K. H. Rehren, and B. Schroer, *Superselection sectors with braid group statistics and exchange algebras, I: General Theory*, Comm. Math. Phys. **125** (1989), 201–226.

[Fro] J. Fröhlich, *Statistics of fields, the Yang-Baxter equation, and the theory of knots and links*, in "The 1987 Cargèse lectures. Non-perturbative quantum field theory," Plenum Press, New York, 1988.

[FGM1] J. Fröhlich, F. Gabbiani and P.-A. Marchetti, *Superselection structure and statistics in three-dimensional local quantum theory*, in "Proceedings 12th John Hopkins Workshop on Current Problems in High Energy Particle Theory Florence 1989" (to appear).

[FGM2] J. Fröhlich, F. Gabbiani and P.-A. Marchetti, *Braid statistics in three-dimensional local quantum theory*, preprint.

[FK] J. Frölich and C. King, *Two-dimensional conformal field theory and three-dimensional topology*, J. Mod. Phys. A (to appear).

[GJ] J. Glimm and A. Jaffe, "Quantum Physics, 2nd ed.," Springer-Verlag, New York Berlin Heidelberg, 1987.

[HK] R. Haag and D. Kastler, *An algebraic approach to quantum field theory*, Comm Math. Phys. **5** (1964), 848–861.

[Ji] M. Jimbo, *Introduction to the Yang-Baxter equation*, preprint.

[Jo] V. K. F. Jones, *Hecke algebra representations of braid groups and link polynomials*, Ann. Math. **126** (1987), 335–388.

[L1] R. Longo, *Index of subfactors and statistics of quantum fields*, Comm. Math. Phys. (to appear).

[L2] R. Longo, *Index of subfactors and statistics of quantum fields II. Correspondences, braid group statistics and Jones polynomial*, preprint.

[ML] S. Mac Lane, "Categories for the Working Mathematician," Springer-Verlag, Berlin, Heidelberg, New York, 1971.

[Ma1] S. Majid, *Quasitriangular Hopf algebras and Yang-Baxter equations*, Int. J. Mod. Phys. A (to appear).

[Ma2] S. Majid, *Quantum group duality in vertex models and other results in the theory of quaistriangular Hopf algebras*, in "Proc. Int. Conf. Geom. Phys., Tahoe City, 1989" (to appear).

[MS] G. Moore and N. Seiberg, *Classical and quantum conformal field theory*, Comm. Math. Phys. **123** (1989), 117–254.

[O] A. Ocneanu, *Quantized groups, string algebras and Galois theory for algebras*.

[S1] B. Schroer, *New concepts and results in nonperturbative quantum field theory*, in "Differential Geometric Methods in Theoretical Physics," Chester, 1988.

[S2] B. Schroer, *High T_C superconductivity and a new symmetry concept in low-dimensional quantum field theories*, preprint.

[S3] B. Schroer, *New kinematics (statistics and symmetry) in low-dimensional QFT with applications to conformal QFT_2*, in "Proceedings of the XVII International Conference on Differential Geometric Methods in Theoretical Physics: Physics and Geometry, 1989" (to appear).

[Su] V. S. Sunder, II_1 *factors, their bimodules and hypergroups*, preprint.

[U] K.-H. Ulbrich, *Tannakian categories for non-commutative Hopf algebras*, preprint.

[Wa] Y. Watatani, *Index for C^*-subalgebras*, Amer. Math. Soc. Memoirs **424** (1990).

[W1] S. L. Woronowicz, *Twisted $SU(2)$. An example of a noncommutative differential calculus*, Publ. Res. Inst. Math. Sci. Kyoto **23** (1987), 117–181.

[W2] S. L. Woronowicz, *Compact matrix pseudogroups*, Comm. Math. Phys. **111** (1987), 613–665.

[W3] S. L. Woronowicz, *Tannaka-Krein duality for compact matrix pseudogroups. Twisted $SU(N)$ groups*, Invent. math. **93** (1988), 35–76.

[W4] S. L. Woronowicz, *Differential calculus on compact matrix pseudogroups (quantum groups)*, Comm. Math. Phys. **122** (1989), 125–170.

Department of Mathematics, University of California, Berkeley CA 94720

Email: rieffel@math.berkeley.edu

Quantum Groups: An Introduction and Survey for Ring Theorists

S. P. SMITH

Abstract. This article describes recent work on new classes of non-commutative algebras which have been dubbed quantum algebras, quantum groups, and quantized enveloping algebras. The basic ring theoretic properties are described, and a number of questions and problems are raised.

A quantum group is a Hopf algebra which is neither commutative, nor co-commutative. Although some authors take this to be the defining property, for the purposes of this introduction, it is permissible to think of a quantum group as a (non-commutative, and non-co-commutative) Hopf algebra which is a deformation of $\mathcal{O}(G)$, the coordinate ring of an affine algebraic group. The deformation depends on a parameter $q \in \mathbb{C}$, such that at $q = 1$ one retrieves $\mathcal{O}(G)$ with its usual Hopf algebra structure. In particular, a quantum group is not a group; in fact, there is no geometric set of points with a binary operation. The relevant point of view is that of non-commutative geometry (à la A. Connes *et al.*), although our approach is that of algebraic rather than differential geometry. Thus, one thinks of (and treats) a non-commutative algebra as if it were the ring of functions on some "non-commutative space" (a "quantum space"). From this point of view a quantum group plays the role of the symmetry group of a quantum space. Hence the subject gives new examples of non-commutative algebras which are modelled on (i.e. are deformations of) classical commutative algebras such as polynomial rings, and regular functions on algebraic groups and related homogeneous spaces.

The use of the adjective "quantum" in this context has two sources. Firstly, quantum groups arose (somewhat indirectly) from a quantum mechanical problem in statistical mechanics (the quantum Yang-Baxter equation); the classical Yang-Baxter equation required for its solution the representation theory of certain classical groups (or more precisely of certain semisimple Lie algebras), and likewise the representation theory of quantum groups plays a role in the solution of the quantum Yang-Baxter equation. Secondly "quantum" evokes the idea of "quantization" which from a mathematical point of view involves the construction of non-commutative algebras which are deformations of certain commutative algebras arising in classical mechanics. The standard example is the Weyl algebra $\mathcal{D}(\mathbf{C}^n)$, the ring of differential operators on $\mathbf{C}^n$, which is the quantization of $\mathcal{O}(\mathbf{C}^{2n})$ arising from the standard Poisson bracket (symplectic structure) on $\mathbf{C}^{2n}$. Thus the adjective "quantum" is used as a heuristic device evoking the origin of the subject, and its relation to quantization.

Quantum groups arose from statistical mechanics via the (quantum) Yang-Baxter equation (QYBE). It was realised around 1982-1983 that solutions to QYBE were intimately related to the finite dimensional modules over certain non-commutative algebras [J1], [J2], [J3]. These non-commutative algebras were completely new objects; while some are sufficiently similar to enveloping algebras of Lie algebras to be quite tractable, many others exhibit new features, offering rich and virgin territory for algebraists. Thus, the first "quantum group" which appeared was not a deformation of a commutative algebra, but a deformation of the enveloping algebra $U\big(sl(n)\big)$, denoted $U_q\big(sl(n)\big)$. It is now possible to first define the deformation $\mathcal{O}_q\big(SL(n)\big)$ of $\mathcal{O}\big(SL(n)\big)$, and then to define $U_q\big(sl(n)\big)$ in terms of $\mathcal{O}_q\big(SL(n)\big)$. It seems that $\mathcal{O}_q\big(SL(n)\big)$ is the more fundamental/natural object. Nevertheless C. M. Ringel [Ri] has recently shown that there is a natural definition of (a subalgebra of) $U_q\big(sl(n)\big)$ in terms of the representation theory of a certain finite dimensional algebra of finite representation type.

The representation theory of quantum groups is similar to that of the ordinary groups. This is not quite so true when q is a root of unity; if q is a root of unity the quantum group is a more complicated object, and these complications are similar to those encountered in the representation theory of semi-simple algebraic groups in positive characteristic (even though the quantum group is in characteristic zero). Hence the interest of representation theorists [An], [DD], [L1], [L2], [L3], [PW1], [PW2], [Xi]. Lusztig has conjectured that the truth of his conjectures on modular representations will follow from the truth of analogous conjectures he has made concerning the representations of quantum groups when the parameter q is a root of unity. There are also close connections with the Hecke algebra, and indeed the Hecke algebra for the symmetric group S_n plays the same role with respect to "quantum $GL(n)$" that the group algebra $\mathbf{C}S_n$ plays with respect to $GL(n)$ viz. the mutual commutant theorem.

We now give a brief description of the contents of this survey.

Section 1 recalls some basic facts about Hopf algebras; in particular the coordinate ring of an algebraic group, and the enveloping algebra of a Lie algebra. Section 2 describes the quantum plane, quantum $\mathbb{P}^1$, quantum $SL(2)$, the quantum exterior algebra, the quantized deformation of

$U(sl(2))$, and the relationships between these objects. Section 3 repeats much of Section 2, but for n-dimensional space. For example, the ring of regular functions on $SL(n, \mathbb{C})$, namely $\mathcal{O}(SL(n))$, is deformed to give a non-commutative and non-cocommutative Hopf algebra, denoted $\mathcal{O}_q(SL(n))$, which depends on a parameter $0 \neq q \in \mathbb{C}$. The words "deformed" and "deformation" will be used informally as a heuristic device; the formalisation involves the fact that at $q = 1$, $\mathcal{O}_q(SL(n)) \cong \mathcal{O}(SL(n))$. Next we define $U_q(\mathfrak{g})$, the deformation of $U(\mathfrak{g})$ for $\mathfrak{g}$ an arbitrary semisimple Lie algebra, and describe its representation theory. Section 4 gives a more abstract approach to the material in Section 3, and in particular for an arbitrary $n^2 \times n^2$ matrix R gives a general construction of a quantum n-space; this is just a non-commutative algebra on n generators where the n^2 defining (quadratic) relations are determined by the matrix R. We also make some remarks on the Yang-Baxter equation. Section 5 describes Manin's approach to quantum groups in the context of quadratic algebras. Some remarks on quadratic algebras and Koszul algebras are made. Section 6 discusses the regular algebras of Artin-Schelter, and Sklyanin algebras. While these are not quantum groups they do appear to be close relatives. Section 7 briefly describes Woronowicz's approach to quantum groups in the context of compact groups and their C^*-algebras. Section 8 describes what happens when q is a root of unity. Finally Section 9 gives some questions/problems that might be usefully addressed by ring theorists.

I have tried to include in the references those papers on quantum groups which are most relevant for ring theorists. These notes have drawn extensively on a number of those articles. In particular, Manin's Montréal notes [M2], Majid [Maj1], and Takhtajan [Tak1]. Moreover, many of the research papers in the area have a significant amount of helpful expository material which I have freely used. I have not reported on that work on quantum groups which falls outside my own area of expertise. In particular, the connections with braids, links and knots, the C^*-algebra aspects, and the physical meaning of the Yang-Baxter equation are all ignored.

I would like to thank T. J. Hodges, M. P. Holland and J. T. Stafford for many helpful discussions concerning the material in this article. I also thank all those who have sent me preprints of their work. New results on quantum groups are appearing at such a rapid rate that this survey is certain to be out of date by the time it appears in print.

§1. Algebraic groups and Hopf algebras

DEFINITION: Let K be a field, and A a K-algebra. Call A a *Hopf algebra* if there are K-algebra homomorphisms $\Delta : A \to A \otimes A$, and $\varepsilon : A \to K$, and a K-algebra anti-homomorphism $s : A \to A$ such that $(\Delta \otimes Id) \circ \Delta = (Id \otimes \Delta) \circ \Delta$, $(Id \bar{\otimes} \varepsilon) \circ \Delta = (\varepsilon \bar{\otimes} Id) \circ \Delta = Id$, and $(s \bar{\otimes} Id) \circ \Delta = (Id \bar{\otimes} s) \circ \Delta = \varepsilon$ (the notation $\theta \bar{\otimes} \varphi$ denotes the map $a \otimes b \mapsto \theta(a)\varphi(b)$). The maps Δ, ε and s are called the *co-product*, the *co-unit* and the *antipode*, respectively.

If A is a K-algebra endowed with maps Δ and ε satisfying the above axioms, then A is called a *bialgebra*. If A is just a vector space endowed with such maps Δ and ε, then A is called a *coalgebra*.

EXAMPLE: Let G be a group, and $\mathbb{C}G$ the group algebra. Define $\Delta(g) = g \otimes g, \varepsilon(g) = 1, s(g) = g^{-1}$ and extend these to linear maps on $\mathbb{C}G$. This gives $\mathbb{C}G$ the structure of a Hopf algebra.

EXAMPLE: Let $\{X_{ij} | 1 \leq i, j \leq n\}$ be the obvious coordinate functions on $M_n(\mathbb{C})$, the ring of complex $n \times n$ matrices. The multiplication map $\mu : M_n(\mathbb{C}) \times M_n(\mathbb{C}) \to M_n(\mathbb{C})$ is a morphism of algebraic varieties, since the ij^{th} entry of the product AB, namely $\sum a_{ik}b_{kj}$, is a polynomial function of the entries in A and B. Hence the comorphism

$$\Delta = \mu^* : \mathcal{O}(M_n) \to \mathcal{O}(M_n \times M_n) \cong \mathcal{O}(M_n) \otimes \mathcal{O}(M_n)$$

is an algebra homomorphism. Because μ is an associative map, it follows that $(\Delta \otimes Id) \circ \Delta = (Id \otimes \Delta) \circ \Delta$. Define $\varepsilon : \mathcal{O}(M_n) \to \mathbb{C}$ to be evaluation at the identity matrix, I. Since the maps $M_n \to M_n \times M_n \to M_n$ given by $A \mapsto (A, I) \mapsto AI$ and $A \mapsto (I, A) \mapsto IA$, are the identity, it follows that $(Id \bar{\otimes} \varepsilon) \circ \Delta = (\varepsilon \bar{\otimes} Id) \circ \Delta = Id$. Thus $\mathcal{O}(M_n)$ with its algebra structure, and the maps Δ and ε, give $\mathcal{O}(M_n)$ the structure of a *bi-algebra*. The data Δ and ε alone gives $\mathcal{O}(M_n)$ the structure of a *co-algebra*. Explicitly, $\Delta(X_{ij}) = \sum_k X_{ik} \otimes X_{kj}$ and $\varepsilon(X_{ij}) = \delta_{ij}$.

Let G be a group which is also an affine algebraic variety over $\mathbb{C}$. If the multiplication map $\mu : G \times G \to G$, and the inverse $g \mapsto g^{-1}$, are both morphisms (of varieties) we call G an *affine algebraic group*. The comorphisms associated to these maps give algebra homomorphisms

$$\Delta : \mathcal{O}(G) \to \mathcal{O}(G) \otimes \mathcal{O}(G)$$
$$s : \mathcal{O}(G) \to \mathcal{O}(G).$$

Define $\varepsilon : \mathcal{O}(G) \to \mathbb{C}$, $f \mapsto f(e)$, to be evaluation at the identity element $e \in G$. As for $M_n(\mathbb{C})$, the group axioms imply relations among Δ, s, and ε which give $\mathcal{O}(G)$ the structure of a Hopf algebra.

EXAMPLE: Let $G = GL(n)$. Since G is the complement of the hypersurface "determinant=0" in $M_n(\mathbb{C})$, $\mathcal{O}(G)$ is a localisation of $\mathcal{O}(M_n) = \mathbb{C}[X_{ij}|1 \leq i,j \leq n]$. Thus $\mathcal{O}(G) = \mathbb{C}[X_{ij}][D^{-1}]$ where $D = \det(X_{ij})$. Thus Δ and ε are exactly as for $M_n(\mathbb{C})$, namely $\Delta(X_{ij}) = \sum_k X_{ik} \otimes X_{kj}$ and $\varepsilon(X_{ij}) = \delta_{ij}$. Furthermore, $s(X_{ij}) = D^{-1}(-1)^{i+j} A_{ji}$ where Λ_{ij} is the ij^{th} $(n-1) \times (n-1)$ minor of the generic matrix (X_{ij}); succinctly, $X s(X) = s(X) X = 1$, where $X = (X_{ij})$.

Every affine algebraic group is isomorphic to a closed subgroup of some $GL(n)$. Thus an arbitrary such group G satisfies $\mathcal{O}(G) \cong \mathcal{O}(GL(n))/I$ for a suitable ideal I. The maps Δ, s, ε defined for $GL(n)$ all pass to the quotient, and induce the correct Hopf algebra structure on $\mathcal{O}(G)$.

The group G can be recaptured from the data in $\mathcal{O}(G)$. The points of G are the maximal spectrum of $\mathcal{O}(G)$, the identity element is the distinguished maximal ideal $\ker(\varepsilon)$, the multiplication is obtained from Δ via $\mu(\mathfrak{m}, \mathfrak{n}) = \ker(\mathcal{O} \xrightarrow{\Delta} \mathcal{O} \otimes \mathcal{O} \to \mathcal{O}/\mathfrak{m} \otimes \mathcal{O}/\mathfrak{n} \cong \mathbb{C})$, the inverse of $\mathfrak{m}$ is $s(\mathfrak{m})$ etc. Since the group G can be discarded, its role usurped by $\mathcal{O}(G)$ with its Hopf algebra structure, one can in some sense justify the attitude that if $\mathcal{O}_q(G)$ is a non-commutative Hopf algebra which is a deformation of $\mathcal{O}(G)$, then $\mathcal{O}_q(G)$ is "functions on a non-commutative space which is a group, that is a quantum group". This attitude is made rather more explicit in [D4] and [M2].

DEFINITION: If A is a Hopf algebra then a left (resp. right) comodule for A is a vector space V together with a map $\lambda : V \to A \otimes V$ (resp. $\rho : V \to V \otimes A$) such that $(Id \otimes \lambda) \circ \lambda = (\Delta \otimes Id) \circ \lambda$ and $(\varepsilon \bar{\otimes} Id) \circ \lambda = Id$ (resp. $(\rho \otimes Id) \circ \rho = (Id \otimes \Delta) \circ \rho$ and $(Id \bar{\otimes} \varepsilon) \circ \rho = Id$).

The representation theory of G can be studied solely through $\mathcal{O}(G)$ in terms of comodules. Let V be a finite dimensional representation of G. If the map $G \to GL(V)$ is a morphism of varieties then V is called a *rational representation*. These are the only representations we shall consider. In that case there is a map $\rho : V \to \mathcal{O}(G) \otimes V$ defined as follows: think of $\mathcal{O}(G) \otimes V$ as functions on G taking values in V; then $\rho(v)(g) = g.v$. The fact that V is a representation ensures that V is an $\mathcal{O}(G)$-comodule. The category of finite dimensional rational G-modules is equivalent to the category of finite dimensional $\mathcal{O}(G)$-comodules (with the appropriate morphisms).

Notation. We shall adopt the conventions of Sweedler. Thus, if A is a Hopf algebra with coproduct Δ, then $\Delta(a) = \sum_{(a)} a_{(1)} \otimes a_{(2)}$, and $(\Delta \otimes Id) \circ \Delta(a) = \sum_{(a)} a_{(1)} \otimes a_{(2)} \otimes a_{(3)}$.

Automorphisms. If V is a rational representation of G, then $S(V)$, the symmetric algebra on V, inherits an action of G as automorphisms. This makes $S(V)$ into a comodule, and the comodule action map $\rho : S(V) \to$

$\mathcal{O}(G) \otimes S(V)$ is an algebra homomorphism. There is also a comodule map $\Lambda(V) \rightarrow \mathcal{O}(G) \otimes \Lambda(V)$ on the exterior algebra (this is also an algebra homomorphism). A simple example is given by $G = GL(n)$, and $V = \mathbb{C}^n = \mathbb{C}X_1 \oplus \ldots \oplus \mathbb{C}X_n$, with the $GL(n)$-action by left multiplication on column vectors. Then $S(V) = \mathbb{C}[X_1, \ldots, X_n]$, and $\rho(X_i) = \sum_k X_{ki} \otimes X_k$.

Notation. If A is a Hopf algebra we shall consider various duals $A^* \supset A^\circ \supset A'$ where

$$A^* = \mathrm{Hom}_{\mathbb{C}}(A, \mathbb{C})$$
$$A^\circ = \{f \in A^* \mid \ker(f) \text{ contains an ideal of finite codimension}\}$$
$$A' = \{f \in A^* \mid \ker(f) \text{ contains } \mathfrak{m}^n \text{ for some } n, \text{ where } \mathfrak{m} = \ker(\varepsilon)\}.$$

Duality. Let G be as above. Write $\mathfrak{m} = \ker(\varepsilon)$. Define

$$U = \{f \in \mathcal{O}(G)^* \mid f(\mathfrak{m}^n) = 0 \text{ for some } n >> 0\}.$$

Thus U is the "continuous dual" of $\mathcal{O}(G)$: U consists of those linear functionals on $\mathcal{O}(G)$ which are continuous in the $\mathfrak{m}$-adic topology. The coalgebra structure on $\mathcal{O}(G)$ transfers to an algebra structure on U: the multiplication in U is $(u.v)(f) = \sum_{(f)} u(f_{(1)}).v(f_{(2)})$ for all $u, v \in U$ and $f \in \mathcal{O}(G)$, and the identity element in U is the co-unit ε. If G is finite, then $\mathcal{O}(G)^\circ \cong \mathbb{C}G$ the group algebra.

If G is connected, and $\mathfrak{g}$ is the Lie algebra of G, defined as the right invariant vector fields (derivations) on G, then $U \cong U(\mathfrak{g})$, the enveloping algebra of $\mathfrak{g}$. The algebra structure on $\mathcal{O}(G)$ transfers to a coalgebra structure on U. In fact, U becomes a Hopf algebra. The maps are $\Delta(X) = X \otimes 1 + 1 \otimes X$, $s(X) = -X$, $\varepsilon(X) = 0$ for all $X \in \mathfrak{g}$. In certain cases $\mathcal{O}(G)$ can be recaptured from $U(\mathfrak{g})$ with its Hopf algebra structure. If $\mathfrak{g}$ is semisimple, then $U(\mathfrak{g})^\circ \cong \mathcal{O}(G)$ where G is the simply connected, connected algebraic group with $\mathrm{Lie}G = \mathfrak{g}$.

Summary. If G is an affine algebraic group then $\mathcal{O}(G)$ is a Hopf algebra, and the representations of G are the same things as $\mathcal{O}(G)$-comodules. Moreover, G and its representation theory may be studied just through $\mathcal{O}(G)$. The enveloping algebra of $\mathfrak{g} = \mathrm{Lie}G$ may be defined in terms of $\mathcal{O}(G)$, through duality. This last fact is of importance in quantum groups because, although $U_q(\mathfrak{g})$ was defined before $\mathcal{O}_q(G)$, it is now clear that one can give a rather more natural definition of $\mathcal{O}_q(G)$, and then define $U_q(\mathfrak{g})$ in terms of $\mathcal{O}_q(G)$ using an analogue of the above duality.

§2. Quantum analogues of $\mathcal{O}(\mathbb{C}^2)$, $\mathcal{O}(SL(2))$ and $U(sl(2))$

This section presents special cases of constructions which will later be given in more generality.

The "algebraic" part of algebraic geometry involves the ring $\mathcal{O}(X)$ consisting of the regular functions on a variety X. The viewpoint of non-commutative geometry is that a non-commutative version of $\mathcal{O}(X)$ consists of "functions on a non-commutative space".

Our discussion of the first example is motivated by the fundamental idea of Dixmier, that primitive ideals in an enveloping algebra $U(\mathfrak{g})$ should be classified in terms of subvarieties (co-adjoint orbits) of $\mathfrak{g}^*$.

EXAMPLE: Fix $0 \neq q \in \mathbb{C}$. The *quantum plane* (more precisely, the coordinate ring of the quantum plane) is the algebra $\mathbb{C}_q[X, Y]$ defined by the relation $YX = q^{-2}XY$. This terminology first appeared in [M1], and has become standard since then. The algebra $\mathbb{C}_q[X, Y]$ has many properties in common with the commutative polynomial ring; *viz.* it is a noetherian domain of global dimension 2, and has a $\mathbb{C}$-basis $\{X^i Y^j | i, j \geq 0\}$. The ways in which $\mathbb{C}_q[X, Y]$ differs from $\mathcal{O}(\mathbb{C}^2)$ depend decisively on whether or not q is a root of unity.

If q is not a root of unity, the only prime ideals in $\mathbb{C}_q[X, Y]$ are $\langle 0 \rangle$, $\langle X \rangle$, $\langle Y \rangle$, $\langle X - \alpha, Y \rangle$, $\langle X, Y - \beta \rangle$ where α and β may be any scalars. Thus every finite dimensional simple module is of dimension 1. There also exist infinite dimensional simple modules, and $\langle 0 \rangle$ is a primitive ideal. All the prime ideals except $\langle X \rangle$ and $\langle Y \rangle$ are primitive. The primitive ideals of finite codimension are in bijection with the points on the X and Y axes in $\mathbb{C}^2$, and the primitive ideal $\langle 0 \rangle$ corresponds to the complement of the axes; in particular, $\mathbb{C}^2$ can be written as a disjoint union of quasi-affine varieties in bijection with the primitive spectrum. We may think of $\mathrm{Spec}\,\mathbb{C}_q[X, Y]$ as consisting of the following irreducible closed sets in $\mathbb{C}^2$: the points on the axes, the axes themselves, $\mathbb{C}^2$ and the empty set $\emptyset$.

Suppose that q^2 is a primitive n^{th} root of 1. Then the center of $\mathbb{C}_q[X, Y]$ is $\mathbb{C}[X^n, Y^n]$, and $\mathbb{C}_q[X, Y]$ is a finitely generated module over its center. Every simple $\mathbb{C}_q[X, Y]$-module is finite dimensional, and there exist simple modules both of dimension 1, and of dimension n. Furthermore $\mathbb{C}_q[X, Y, X^{-1}, Y^{-1}]$ is an Azumaya algebra of rank n^2 over its center.

There is a bialgebra structure on $\mathbb{C}_q[X, Y]$ given by $\Delta(X) = X \otimes X$, $\Delta(Y) = Y \otimes 1 + X \otimes Y$, $\varepsilon(X) = 1$, $\varepsilon(Y) = 0$.

REMARK: To what extent is $\mathbb{C}_q[X, Y]$ "natural"? Why call this the quantum plane rather than some other nice algebra of GK-dimension 2? If one wants an algebra $\mathbb{C}[X, Y]$ with basis $X^i Y^j$, then the only possibilities (up to isomorphism of algebras) are those defined by one of the following relations:

(a) $YX = 0$ (b) $XY - YX = 1$

(c) $XY - \lambda YX = 1$ ($\lambda \in \mathbb{C}^*$)   (d) $XY = \lambda YX$ ($\lambda \in \mathbb{C}^*$)

(e) $XY - YX = Y$ (f) $XY - YX = Y^2$

Hence if one further requires a graded algebra with X and Y of degree 1, then the only possibilities are (a), (d) and (f). In a sense (which can be made precise) (d) is the generic example. Thus $\mathbb{C}_q[X, Y]$ is not at all arbitrary.

The coordinate ring of quantum 2×2 matrices. This is the algebra generated by elements a, b, c, d subject to the relations

$$ba = q^{-2}ab \qquad ca = q^{-2}ac \qquad\qquad bc = cb$$
$$db = q^{-2}bd \qquad dc = q^{-2}cd \qquad\qquad ad - da = (q^2 - q^{-2})bc$$

We will write $\mathcal{O}_q(M_2(\mathbb{C})) := \mathbb{C}[a, b, c, d] = \mathbb{C}\begin{bmatrix} a & b \\ c & d \end{bmatrix}$. This is an iterated Ore extension, which means, in this situation, that there is a sequence of subalgebras

$$\mathbb{C}[a] \subset \mathbb{C}[a, b] = \bigoplus_j \mathbb{C}[a]b^j$$

$$\subset \mathbb{C}[a, b, c] = \bigoplus_k \mathbb{C}[a, b]c^k$$

$$\subset \mathcal{O}_q(M_2(\mathbb{C})) = \bigoplus_\ell \mathbb{C}[a, b, c]d^\ell.$$

Hence $\mathcal{O}_q(M_2(\mathbb{C}))$ is a noetherian domain with basis $\{a^i b^j c^k d^\ell \mid (i, j, k, \ell) \in \mathbb{N}^4\}$. The basis can be obtained directly from the defining relations by applying the Diamond Lemma [Be]. The center of this ring (if q is not a root of unity) is generated by $ad - q^2 bc = da - q^{-2}bc$. Define a comultiplication and co-unit on $\mathcal{O}_q(M_2(\mathbb{C}))$ by

$$\Delta(a) = a \otimes a + b \otimes c, \quad \Delta(b) = a \otimes b + b \otimes d,$$
$$\Delta(c) = c \otimes a + d \otimes c, \quad \Delta(d) = c \otimes b + d \otimes d$$
$$\varepsilon(a) = \varepsilon(d) = 1, \qquad \varepsilon(b) = \varepsilon(c) = 0.$$

One may check that $\mathcal{O}_q(M_2(\mathbb{C}))$ becomes a bialgebra. The coproduct and co-unit can be easily understood by writing them as $\Delta(X_{ij}) = \sum_k X_{ik} \otimes X_{kj}$, and $\varepsilon(X_{ij}) = \delta_{ij}$, with the notation $a = X_{11}$, $b = X_{12}$, $c = X_{21}$, $d = X_{22}$.

The coordinate ring of quantum $SL(2)$. This is the algebra $\mathcal{O}_q(SL(2)) = \mathcal{O}_q(M_2(\mathbb{C}))/\langle ad - q^2bc - 1\rangle$. This extra relation is the quantum analogue of the relation "determinant=1". We will see later that $\mathcal{O}_q(SL(2))$ can be defined naturally in terms of the quantum plane via a suitable universal property, whence the naturality of the defining relations for $\mathcal{O}_q(M_2(\mathbb{C}))$ and $\mathcal{O}_q(SL(2, \mathbb{C}))$ is a consequence of the naturality of $\mathbb{C}_q[X, Y]$, discussed earlier. Since $\Delta(ad - q^2bc) = (ad - q^2bc) \otimes (ad - q^2bc)$, and $\ker(\varepsilon) \supset \langle ad - q^2bc - 1\rangle$, both Δ and ε pass to the quotient, making $\mathcal{O}_q(SL(2))$ a bialgebra. In addition, there is an antipode on $\mathcal{O}_q(SL(2))$ defined by

$$s\begin{bmatrix} a & b \\ c & d \end{bmatrix} = (ad - q^2bc)^{-1}\begin{bmatrix} d & -q^{-2}b \\ -q^2c & a \end{bmatrix}$$

which is a short way of writing $s(a) = (ad - q^2bc)^{-1}d$ etc. The antipode has finite order if and only if q is a root of unity, and is an involution when $q^4 = 1$. In this way $\mathcal{O}_q(SL(2))$ becomes a Hopf algebra. At $q = 1$ one retrieves $\mathcal{O}(SL(2))$ with its usual structure.

Henceforth suppose that q is not a root of unity.

DEFINITION OF $U_q(sl(2))$: The q-analogue of the enveloping algebra of the Lie algebra $sl(2)$ is $U_q(sl(2)) = \mathbb{C}[E, F, K, K^{-1}]$ defined by relations

$$KE = q^2EK, \qquad KF = q^{-2}FK, \qquad EF - FE = \frac{K^2 - K^{-2}}{q^2 - q^{-2}}.$$

There is a Hopf algebra structure on $U_q(sl(2))$ defined by

$$\Delta(E) = E \otimes K^{-1} + K \otimes E, \qquad s(E) = -q^{-2}E, \qquad \varepsilon(E) = 0$$
$$\Delta(F) = F \otimes K^{-1} + K \otimes F, \qquad s(F) = -q^2F, \qquad \varepsilon(F) = 0$$
$$\Delta(K) = K \otimes K \qquad\qquad\qquad s(K) = K^{-1}, \qquad \varepsilon(K) = 1.$$

Both Δ and ε extend to algebra homomorphisms, and s extends to an algebra anti-homomorphism.

Clearly $U_q(sl(2))$ is an iterated Ore extension, hence a noetherian domain of GK-dimension 3, and has a basis $\{E^iF^jK^\ell \mid (i, j) \in \mathbb{N}^2, \ell \in \mathbb{Z}\}$. The center of $U_q(sl(2))$ is generated by an analogue of the Casimir element,

$$\Omega = EF + FE + \left(\frac{q^4 + 1}{q^4 - 1}\right)\left(\frac{K^2 + K^{-2}}{q^2 - q^{-2}}\right).$$

To understand better the relationship of $U_q(sl(2))$ to $U(sl(2))$, see the remark about deformations at the end of this section.

Representations of $U_q(sl(2))$. For each $n \in \mathbb{N}$, there are precisely 4 non-isomorphic simple $U_q(sl(2))$-modules of dimension n, and every finite dimensional $U_q(sl(2))$-module is semisimple. Thus the representation theory of $U_q(sl(2))$ is remarkably similar to that of $sl(2)$. The four 1-dimensional modules are given by $E \mapsto 0$, $F \mapsto 0$, $K \mapsto \omega$ where $\omega^4 = 1$. One of the two dimensional irreducible representations is given by $\psi : U_q(sl(2)) \to M_2(\mathbb{C})$

$$E \mapsto \begin{bmatrix} 0 & 1 \\ 0 & 0 \end{bmatrix} \qquad F \mapsto \begin{bmatrix} 0 & 0 \\ 1 & 0 \end{bmatrix} \qquad K \mapsto \begin{bmatrix} q & 0 \\ 0 & q^{-1} \end{bmatrix}.$$

The other 2-dimensional representations are obtained from this one as follows: let $\omega^4 = 1$, and in the above three matrices replace 1 by ω, q by ωq, and q^{-1} by ωq^{-1}.

The co-product on $U_q(sl(2))$, allows a tensor product of modules to be given the structure of a $U_q(sl(2))$-module. For example, tensoring this particular 2-dimensional representation, with the four 1-dimensional representations gives all four 2-dimensional representations. Furthermore, these examples illustrate that in general, if V and W are $U_q(sl(2))$-modules then the twist map from $V \otimes W$ to $W \otimes V$ need not be an isomorphism.

It is interesting to tensor the above 2-dimensional representation with itself. First denote the 2-dimensional representation by $V = \mathbb{C}X \oplus \mathbb{C}Y$, where $\{X, Y\}$ is the ordered basis such that $\psi(E), \psi(F)$ and $\psi(K)$ are as above. As in the classical case $V^{\otimes 2}$ is a direct sum of a 1-dimensional and a 3-dimensional simple module, *viz.*

$$\langle X^2, \ qXY + q^{-1}YX, \ Y^2 \rangle \oplus \langle q^{-1}XY - qYX \rangle.$$

In contrast to the classical situation, the symmetric tensors do not form a submodule, and there is no natural action of $U_q(sl(2))$ on the symmetric algebra. However, the role of the symmetric algebra on $\mathbb{C}^2$ is played by $\mathbb{C}_q[X, Y]$, and the analogy with the classical situation will be reinforced by the way in which $\mathbb{C}_q[X, Y]$ may be made into a $U_q(sl(2))$-module. To describe this action succinctly, first recall the notion of a Hopf module algebra.

DEFINITION: If H is a Hopf algebra, and A an algebra which is an H-module in such a way that

(i) $h.(ab) = \sum_{(h)} h_{(1)}(a).h_{(2)}(b)$ \qquad for all $h \in H$, $a, b \in A$

(ii) $h.1 = \varepsilon(h)1$ \quad for all $h \in H$, where $1 \in A$ is the identity, then A is an *H-module algebra*. In particular, if the coproduct on H is known, then to specify the action of H on A it is enough to say how the generators of H act on the generators of A.

EXAMPLES: 1. If an algebra A has an action of a group G as automorphisms, then the extension of this action to $\mathbb{C}G$ makes A a $\mathbb{C}G$-module algebra.

2. If A is a $U(\mathfrak{g})$-module algebra, then each $X \in \mathfrak{g}$ must act as a derivation: the fact that $\Delta(X) = 1 \otimes X + X \otimes 1$ is another way of expressing the product rule for derivations, that is $X(ab) = aX(b) + X(a)b$.

Now make $\mathbb{C}_q[X, Y]$ a $U_q(sl(2))$-module algebra by defining

$$
\begin{aligned}
E(X) &= 0 & F(X) &= q^{-1}Y & K(X) &= qX \\
E(Y) &= qX & F(Y) &= 0 & K(Y) &= q^{-1}Y
\end{aligned}
$$

This is analogous to the action of $U(sl(2))$ as differential operators on the commutative ring (symmetric algebra) $S(\mathbb{C}^2) = \mathbb{C}[X, Y]$ via $E = X\frac{\partial}{\partial Y}$, $H = X\frac{\partial}{\partial X} - Y\frac{\partial}{\partial Y}$, $F = Y\frac{\partial}{\partial X}$. The decomposition of $S(\mathbb{C}^2) = \oplus_n S^n(\mathbb{C}^2)$ into its homogeneous components is an $sl(2)$-module decomposition, and each symmetric power $S^n(\mathbb{C}^2)$ is the unique $(n+1)$-dimensional $sl(2)$-module. Likewise, the degree n homogeneous component of $\mathbb{C}_q[X, Y]$ is a simple $U_q(sl(2))$-module (the simplicity depends on the fact that q is not a root of unity). The complete list of finite dimensional simple $U_q(sl(2))$-modules is obtained by tensoring the homogeneous components of $\mathbb{C}_q[X, Y]$ with the four 1-dimensional modules.

The above construction may be obtained as follows. First, there is a unique way of extending the $U_q(sl(2))$ action on V to an action on $T(V)$, the tensor algebra, such that $T(V) = \mathbb{C}\langle X, Y \rangle$ is a $U_q(sl(2))$-module algebra. Then observe that the ideal $\langle q^{-1}XY - qYX \rangle$ is a $U_q(sl(2))$-submodule. Hence the quotient algebra, which is $\mathbb{C}_q[X, Y]$, becomes a $U_q(sl(2))$-module algebra.

Geometry. Is there any geometry associated to the representation theory of $U_q(sl(2))$? Is there an analogue of the fact that the finite dimensional irreducible representations of $sl(2)$ may be realised as global sections of the line bundles $\mathcal{O}(n), n \geq 0$, on $\mathbb{P}^1$? Recall that this is expressed by the fact that $\oplus_n H^0(\mathbb{P}^1, \mathcal{O}(n)) \cong \mathbb{C}[X, Y]$ the homogeneous coordinate ring of $\mathbb{P}^1$, and the action of $sl(2)$ is that given by the vector fields above. Hence the action of $U_q(sl(2))$ on $\mathbb{C}_q[X, Y]$ is the analogue of this, and the homogeneous components of $\mathbb{C}_q[X, Y]$ are "the global sections of line bundles on quantum $\mathbb{P}^1$"

To make this precise, recall that the homogeneous prime ideals of $\mathbb{C}_q[X, Y]$ are $\langle X, Y \rangle, \langle X \rangle, \langle Y \rangle, \langle 0 \rangle$; ignore the irrelevant ideal $\langle X, Y \rangle$. Hence $\mathrm{Proj}\,\mathbb{C}_q[X, Y]$, the homogeneous spectrum of $\mathbb{C}_q[X, Y]$, consists of 3 points. The closed points form a two point topological space $\{\langle X \rangle, \langle Y \rangle\}$ in which every subset is closed. However, it seems better to take the following approach which is consistent with the earlier description of the prime ideals of $\mathbb{C}_q[X, Y]$ in terms of certain subvarieties of $\mathbb{C}^2$. Consider the usual $\mathbb{P}^1$ but give it a new topology determined by the homogeneous spectrum of $\mathbb{C}_q[X, Y]$. Thus the closed sets, and the corresponding homogeneous radical ($\equiv$semiprime) ideals are

$$
\begin{array}{ccccc}
\mathbb{P}^1, & \emptyset, & \{(0,1)\}, & \{(1,0)\}, & \{(0,1),(1,0)\} \\
\langle 0 \rangle, & \mathbb{C}_q[X,Y], & \langle X \rangle, & \langle Y \rangle, & \langle XY \rangle
\end{array}
$$

The complementary open sets are

$$
\begin{array}{ccccc}
\emptyset, & \mathbb{P}^1, & U_X, & U_Y, & U_{XY}.
\end{array}
$$

This may seem rather little data with which to do "non-commutative algebraic geometry"; however, a quasi-coherent $\mathcal{O}_{\mathbb{P}^1}$-module $\mathcal{M}$ is completely determined by the local sections $\mathcal{M}(U_X)$, $\mathcal{M}(U_Y)$, $\mathcal{M}(U_{XY})$ as modules over $\mathcal{O}(U_X)$, $\mathcal{O}(U_Y)$, $\mathcal{O}(U_{XY})$, and the restriction maps $\mathcal{M}(U_X) \to \mathcal{M}(U_{XY})$ and $\mathcal{M}(U_Y) \to \mathcal{M}(U_{XY})$. We may use $\mathbb{C}_q[X, Y]$ to define the structure sheaf of "quantum $\mathbb{P}^1$". For example, to obtain $\mathcal{O}(U_X)$, first localise to obtain $\mathbb{C}_q[X, Y, X^{-1}]$, and then take the degree 0 part to obtain $\mathbb{C}[Y/X] \cong \mathcal{O}(\mathbb{C})$. It is clear that the structure sheaf of "quantum $\mathbb{P}^1$" is the "same" as the structure sheaf of the usual $\mathbb{P}^1$. Thus quantum $\mathbb{P}^1$ is exactly the same as the usual $\mathbb{P}^1$. This is not too surprising because there are no deformations of $\mathbb{C}[X]$, whence quantum $\mathbb{A}^1$ is the same as the usual $\mathbb{A}^1$. The usual commutative proof shows that there is an equivalence of categories between coherent $\mathcal{O}_{\mathbb{P}^1}$-modules, and finitely generated graded $\mathbb{C}_q[X, Y]$-modules modulo the finite dimensional modules (this is a very special case of a result due to Artin and van den Bergh [AV]).

Duality. It is natural to ask for the relation between $\mathcal{O}_q(SL(2))$, and $U_q(sl(2))$. In contrast to our order of presentation, $U_q(sl(2))$ was the first object to be defined. Recall the classical duality expressing $U(sl(2))$ as the continuous dual of $\mathcal{O}(SL(2))$ with respect to the $\mathfrak{m}$-adic topology, where $\mathfrak{m} = \ker(\varepsilon)$. The strict analogue of this fails: if $q^2 \neq 1$, then $\bigcap \mathfrak{m}^n = \langle b, c \rangle$, and thus $\mathcal{O}_q\big(SL(2)\big)' = \mathcal{O}(T)'$ where T is the 1-dimensional torus with $\mathcal{O}(T) = \mathbb{C}[a,d]$ where $ad = 1$. However, $\mathcal{O}(T)' \cong \mathcal{O}(\mathbb{C})$. Nevertheless, there is an algebra homomorphism $\mathcal{O}_q\big(SL(2)\big) \to U_q\big(sl(2)\big)^*$ due to Rosso [R1] which we now describe. Let $\rho_{11}, \rho_{12}, \rho_{21}, \rho_{22}$ be the usual coordinate functions on $M_2(\mathbb{C})$. By composing the ρ_{ij} with $\psi : U_q\big(sl(2)\big) \to M_2(\mathbb{C})$, the homomorphism determined by the 2-dimensional representation described above, one may view the ρ_{ij} as elements of $U_q\big(sl(2)\big)^*$. Rosso [R1] says that there is an isomorphism $\mathcal{O}_q\big(SL(2)\big) \to \mathbb{C}[\rho_{11}, \rho_{12}, \rho_{21}, \rho_{22}]$.

Nowadays, $\mathcal{O}_q\big(SL(2)\big)$ is viewed as the more fundamental object, and hence it is appropriate to define $U_q\big(sl(2)\big)$ in terms of $\mathcal{O}_q\big(SL(2)\big)$. The dual of the map $\mathcal{O}_q\big(SL(2)\big) \to U_q\big(sl(2)\big)^*$ gives an injective (this is not immediately obvious) map $U_q\big(sl(2)\big) \to \mathcal{O}_q\big(SL(2)\big)^*$, and the image is the full continuous dual with respect to a suitable co-finite topology on $\mathcal{O}_q\big(SL(2)\big)$. Thus $U_q\big(sl(2)\big)$ can be defined to be the continuous dual of $\mathcal{O}_q\big(SL(2)\big)$ with respect to this topology.

Dual to the fact that $\mathbb{C}_q[X,Y]$ is a left $U_q\big(sl(2)\big)$-module, there is an algebra homomorphism $\rho : \mathbb{C}_q[X,Y] \to \mathbb{C}_q[X,Y] \otimes \mathcal{O}_q\big(SL(2)\big)$ making $\mathbb{C}_q[X,Y]$ a right $\mathcal{O}_q\big(SL(2)\big)$-comodule. A calculation shows that $\rho(X) = X \otimes a + Y \otimes c$ and $\rho(Y) = X \otimes b + Y \otimes d$.

The quantum exterior algebra. Recall the decomposition of $V^{\otimes 2}$ as the sum of two $U_q\big(sl(2)\big)$-modules. Just as the 1-dimensional submodule determined an ideal of $T(V)$, with quotient $\mathbb{C}_q[X,Y]$, so does the 3-dimensional submodule determine an ideal. The quotient of $T(V)$ by this ideal is called the *quantum exterior algebra*, and is denoted $\Lambda_q(\mathbb{C}^2)$. To avoid any confusion label the images of X and Y as ξ and η. Hence $\Lambda_q(\mathbb{C}^2) = \mathbb{C}[\xi, \eta]$ with relations $\xi^2 = \eta^2 = q^2\xi\eta + \eta\xi = 0$. By construction, this becomes a left $U_q\big(sl(2)\big)$-module algebra. Similarly, $\Lambda_q(\mathbb{C}^2)$ becomes a right $\mathcal{O}_q\big(SL(2)\big)$-comodule, with $\rho : \Lambda_q(\mathbb{C}^2) \to \Lambda_q(\mathbb{C}^2) \otimes \mathcal{O}_q\big(SL(2)\big)$ given by $\rho(\xi) = \xi \otimes a + \eta \otimes c$, and $\rho(\eta) = \xi \otimes b + \eta \otimes d$. Observe that $\rho(\xi\eta) = \xi\eta \otimes (ad - q^2 bc)$, so that the quantum determinant appears in an analogous way to the classical determinant.

A deformation of $U\big(sl(2)\big)$. When $q = 1, \mathcal{O}_q\big(SL(2)\big) \cong \mathcal{O}\big(SL(2)\big)$, and this allows one to view $\mathcal{O}_q\big(SL(2)\big)$ as a deformation of $\mathcal{O}\big(SL(2)\big)$. Is there an analogue for $U_q\big(sl(2)\big)$ and $U\big(sl(2)\big)$? Since the defining relations of $U_q\big(sl(2)\big)$ do not make sense at $q = 1$, the answer is "no". However, as already mentioned, $U_q\big(sl(2)\big)$ was not the first "analogue" of $U\big(sl(2)\big)$ to be defined.

Let T be the free algebra on E, F, H over the polynomial ring $\mathbb{C}[h]$. Let $\widehat{T}$ be the completion of T with respect to the h-adic topology determined by the ideals $h^n T$. Let $\widehat{I}$ be the closed 2-sided ideal of $\widehat{T}$ generated by

$$HE - EH - 2E, \quad HF - FH + 2F, \quad EF - FE - \sinh(\tfrac{1}{2}hH)/\sinh(\tfrac{1}{2}h).$$

Define $U_h\big(sl(2)\big) := \widehat{T}/\widehat{I}$; this is the algebra originally defined by Kulish and Reshetikhin [KR]. In particular, $U_h\big(sl(2)\big)/hU_h\big(sl(2)\big) \cong U\big(sl(2)\big)$. To see this, expand $f(h) = \sinh(\tfrac{1}{2}hH)/\sinh(\tfrac{1}{2}h)$ as a power series in h. Although $f(0)$ is not defined, l'Hôpital's rule gives $\lim_{h\to 0} f(h) = H$, hence the power series begins $f(h) = H + \cdots$.

To understand the relation between $U_h\big(sl(2)\big)$ and $U_q\big(sl(2)\big)$, treat h as a complex number and set $q = \exp(\tfrac{1}{4}h) = e^{h/4}$, and $K = \exp(\tfrac{1}{4}hH)$. Now the third relation in $U_h\big(sl(2)\big)$ becomes $EF - FE = (K^2 - K^{-2})/(q^2 - q^{-2})$. Furthermore, if $g(H) \in \mathbb{C}[H]$, then $g(H)E = Eg(H + 2)$ and $g(H)F = Fg(H - 2)$. Hence $KE = q^2 EK$ and $KF = q^{-2}FK$. Thus the subalgebra $\mathbb{C}[E, F, K, K^{-1}]$ is isomorphic to $U_q\big(sl(2)\big)$.

For every semisimple Lie algebra $\mathfrak{g}$, there is such an algebra $U_h(\mathfrak{g})$. These algebras are studied in detail in two papers of Tanisaki, ([Tan1], [Tan2]), and by Rosso [R5].

§3. Quantum n-space, $\mathcal{O}_q\big(SL(n)\big)$ and $U_q\big(sl(n)\big)$

This section discusses the appropriate higher dimensional analogues of the quantum objects described in the previous section.

Quantum affine n-space. Let $\mathcal{O}_q(\mathbb{C}^n) = \mathbb{C}[X_1, \dots, X_n]$ be the algebra defined by the relations $X_j X_i = q^{-2} X_i X_j$ whenever $i < j$. Call $\mathcal{O}_q(\mathbb{C}^n)$ the coordinate ring of quantum n-space, or just *quantum affine n-space*. The basic ring theoretic properties of $\mathcal{O}_q(\mathbb{C}^n)$ are easy to establish. For example, it is a noetherian domain of GK-dimension n, and has a basis given by the monomials X^I where $I = (i_1, \dots, i_n)$ is a multi-index with each $i_j \geq 0$. The homogeneous prime ideals are the ideals of the form

$$\langle Y_1, \dots, Y_k \rangle \text{ where } \{Y_1, \dots, Y_k\} \subset \{X_1, \dots, X_n\}.$$

In particular, the homogeneous primes correspond to intersections of the coordinate hyperplanes in $\mathbb{C}^n$. All the homogeneous prime ideals are sums of the height 1 primes $\langle X_i \rangle$, so as in the previous section, one can think of $\mathrm{Proj}\mathcal{O}_q(\mathbb{C}^n)$ (quantum $\mathbb{P}^{n-1}$) being $\mathbb{P}^{n-1}$ with a new topology (the quantum Zariski topology). In this topology the basic open sets are the standard open sets $U_i = \{(\alpha_1, \dots, \alpha_n)|\alpha_i \neq 0\}$. The structure sheaf on quantum $\mathbb{P}^{n-1}$ has as sections over U_i the degree zero part of $\mathcal{O}_q(\mathbb{C}^n)[X_i^{-1}]$, namely

$\mathbb{C}[X_1 X_i^{-1}, \dots, X_n X_i^{-1}]$. This is isomorphic to $\mathcal{O}_q(\mathbb{C}^{n-1})$, so U_i with its sections of the structure sheaf is isomorphic to quantum affine $(n-1)$-space. Also, note that quantum $\mathbb{P}^{n-1}$ is an $(n-1)$-dimensional topological space.

Determining the non-homogeneous prime ideals of $\mathcal{O}_q(\mathbb{C}^n)$ is a little trickier. By [MP, Proposition 1.3], $\mathcal{O}_q(\mathbb{C}^n)[X_1^{-1}, \dots, X_n^{-1}]$ is simple if n is even. Hence, if n is even, a non-zero prime ideal contains some X_i. Since $\mathcal{O}_q(\mathbb{C}^n)/\langle X_i \rangle \simeq \mathcal{O}_q(\mathbb{C}^{n-1})$, the question becomes that of describing $\operatorname{Spec}\mathcal{O}_q(\mathbb{C}^n)$ with n odd.

Suppose n is odd. Now [MP, Proposition 1.3] says that $\mathcal{O}_q(\mathbb{C}^n)[X_1^{-1}, \dots, X_n^{-1}]$ is not simple. In fact, the normal element $t := X_1 X_3 X_5 \cdots X_n - X_2 X_4 \cdots X_{n-1}$, generates a proper prime ideal. This was pointed out to me by A. Bell. The minimal primitive ideals of $\mathcal{O}_q(\mathbb{C}^n)$ (which are also the minimal non-zero prime ideals) are the $\langle X_i \rangle$, $1 \leq i \leq n$, and the ideals

$$J_{\alpha,\beta} = \langle \alpha X_1 X_3 \cdots X_n - \beta X_2 X_4 \cdots X_{n-1} \rangle$$

for $\alpha, \beta \in \mathbb{C}$ with $\alpha\beta \neq 0$. Any prime $P \not\supseteq J_{\alpha,\beta}$ must contain $\langle X_i, X_j \rangle$ for some i odd, j even. Hence this describes all the height one and height two primes, and if P is a height two prime then $\mathcal{O}_q(\mathbb{C}^n)/P \cong \mathcal{O}_q(\mathbb{C}^{n-2})$. Thus induction will yield all the prime ideals.

Quantum matrices, quantum $SL(n)$ and quantum $GL(n)$. Define $\mathcal{O}_q(M_n(\mathbb{C})) = \mathbb{C}[X_{ij} \mid 1 \leq i,j \leq n]$ subject to the relations that if $i < j$ and $k < m$, the map $\mathcal{O}_q(M_2(\mathbb{C})) \to \mathbb{C}[X_{ik}, X_{im}, X_{jk}, X_{jm}]$ given by

$$\begin{bmatrix} a & b \\ c & d \end{bmatrix} \longrightarrow \begin{bmatrix} X_{ik} & X_{im} \\ X_{jk} & X_{jm} \end{bmatrix}$$

is a ring isomorphism. It is quite easy to see that $\mathcal{O}_q(M_2(\mathbb{C}))$ is an iterated Ore extension: just adjoin the variables in the order $X_{11}, X_{12}, X_{21}, X_{22}$, $X_{13}, X_{31}, X_{23}, X_{32}, X_{33}, X_{14}, \dots$ etc. Therefore $\mathcal{O}_q(M_n(\mathbb{C}))$ is a noetherian domain.

Write S_n for the symmetric group on n letters, and write $\ell(\sigma)$ for the length of $\sigma \in S_n$; thus $\ell(\sigma)$ is the minimal number of terms required to express σ as a product of the simple transpositions $(i, i+1)$. The *quantum determinant* is the element

$$\det{}_q X = \sum_{\sigma \in S_n} (-q^{-2})^{\ell(\sigma)} X_{1,\sigma 1} X_{2,\sigma 2} \cdots X_{n,\sigma n}.$$

The center of $\mathcal{O}_q(M_n(\mathbb{C}))$ is $\mathbb{C}[\det_q X]$ (the center is much larger if q is a root of unity; see §7). This is stated in [FRT]. A proof that $\det_q X$ is central is given in [PW1, Theorem 4.6.1], and a proof that it generates the center is given in [NYM, Theorem 2.2].

Define quantum $SL(n)$ by $\mathcal{O}_q(SL(n)) := \mathcal{O}_q(M_n(\mathbb{C}))/\langle \det_q X - 1 \rangle$, and define quantum $GL(n)$ by $\mathcal{O}_q(GL(n)) := \mathcal{O}_q(M_n(\mathbb{C}))[(\det_q X)^{-1}]$. For $q = 1$, the constructions give $\mathcal{O}(SL(n))$ and $\mathcal{O}(GL(n))$, respectively.

There is a Hopf algebra structure on $\mathcal{O}_q(SL(2))$ (and on $\mathcal{O}_q(GL(n))$). The coproduct is $\Delta(X_{ij}) = \sum_k X_{ik} \otimes X_{kj}$, and the co-unit is $\varepsilon(X_{ij}) = \delta_{ij}$. The explicit description of the antipode, s say, is rather complicated, but is uniquely determined by the requirement that $Xs(X) = I_n = s(X)X$ where $X = (X_{ij})$ is the matrix in $M_n(\mathcal{O}_q(SL(n)))$, $s(X)$ is the matrix $(s(X_{ij}))$, and I_n is the $n \times n$ identity matrix.* Both Δ and ε are algebra homomorphisms, and s is an anti-homomorphism. Notice that $\Delta(\det_q X) = (\det_q X) \otimes (\det_q X)$, and $\varepsilon(\det_q X) = 1$.

We now describe s explicitly. It is clear from the defining relations, that if p rows and p columns are deleted from the generic matrix $X = (X_{ij})$, then the subalgebra of $\mathcal{O}_q(M_n(\mathbb{C}))$ generated by the remaining X_{ij} is isomorphic to $\mathcal{O}_q(M_{n-p}(\mathbb{C}))$. If one deletes row i, and column j, let Y denote the $(n-1) \times (n-1)$ generic matrix so obtained. Set $A_{ji} = \det_q Y \in \mathcal{O}_q(M_{n-1}(\mathbb{C}))$. Then

$$s(X_{ij}) = (-q^{-2})^{j-i} A_{ji}(\det_q X)^{-1}.$$

There is a q-analogue of the action of $GL(n)$ on $\mathbb{C}^n$ viz. $\mathcal{O}_q(\mathbb{C}^n)$ can be made into both a right, and a left $\mathcal{O}_q(GL(n))$-comodule via maps are $\rho(X_i) = \sum_k X_k \otimes X_{ki}$ and $\lambda(X_i) = \sum_k X_{ik} \otimes X_k$.

Universal property of $\mathcal{O}_q(GL(n))$. The definition of $\mathcal{O}_q(GL(n))$ can be made to appear more natural by showing that $\mathcal{O}_q(GL(n))$ satisfies a certain universal property with respect to $\mathcal{O}_q(\mathbb{C}^n)$ (see [M2] for details). Let H be a Hopf algebra, and suppose there are algebra maps $\lambda' : \mathcal{O}_q(\mathbb{C}^n) \to \mathcal{O}_q(\mathbb{C}^n) \otimes H$ and $\rho' : \mathcal{O}_q(\mathbb{C}^n) \to H \otimes \mathcal{O}_q(\mathbb{C}^n)$ making $\mathcal{O}_q(\mathbb{C}^n)$ a left and right comodule, and also such that $\lambda'(X_i) = \sum_k Y_{ik} \otimes X_k$ and $\rho'(X_i) = \sum_k X_k \otimes Y_{ki}$. Then there is a unique algebra homomorphism $\psi : \mathcal{O}_q(GL(n)) \to H$ such that $\lambda' = (\psi \otimes Id) \circ \lambda$ and $\rho' = (Id \otimes \psi) \circ \rho$. This is not quite the same way that Manin [M2] presents the universal property, although it is equivalent because we are working with a distinguished set of generators for $\mathcal{O}_q(GL(n))$. We will return to this point in §5.

Definition of $U_q(\mathfrak{g})$: Let $\mathfrak{g}$ be a complex semisimple Lie algebra of rank n, with Cartan matrix $A = (a_{ij})$. There are integers $d_1, \ldots, d_n \in \{1, 2, 3\}$

---

*The uniqueness of such an $s(X)$ is guaranteed by Goldie's Theorem: $M_n(\mathcal{O}_q(M_n(\mathbb{C})))$ is a prime noetherian ring, so once we know that $X = (X_{ij})$ is regular, the existence and uniqueness of $s(X)$ follows.

such that $d_i a_{ij} = d_j a_{ji}$ i.e. $(d_i a_{ij})$ is a symmetric matrix.

Let t be an indeterminate. Define, for $m > n > 0$

$$[m]_t := \prod_{j=1}^{m} \frac{t^j - t^{-j}}{t - t^{-1}} \in \mathbb{Z}[t, t^{-1}]$$

$$\begin{bmatrix} m \\ n \end{bmatrix}_t := \frac{[m]_t}{[n]_t [m-n]_t}$$

$$= \frac{(t^m - t^{-m})(t^{(m-1)} - t^{-(m-1)}) \cdots (t^{(m-n+1)} - t^{-(m-n+1)})}{(t - t^{-1})(t^2 - t^{-2}) \cdots (t^n - t^{-n})}$$

This is called the t-binomial coefficient. We adopt the conventions that $\begin{bmatrix} m \\ 0 \end{bmatrix}_t = 1$, and $\begin{bmatrix} m \\ n \end{bmatrix}_t = 0$ if $n < 0$ or $n > m$.

Define $U_q(\mathfrak{g}) = \mathbb{C}[E_i, K_i, K_i^{-1}, F_i | 1 \leq i \leq n]$ with relations

$$K_i K_i^{-1} = K_i^{-1} K_i = 1 \qquad K_i K_j - K_j K_i = 0$$

$$E_i F_j - F_j E_i = \delta_{ij} \left(\frac{K_i^2 - K_i^{-2}}{q^{2d_i} - q^{-2d_i}} \right)$$

$$K_i E_j K_i^{-1} = q^{d_i a_{ij}} E_j \qquad K_i F_j K_i^{-1} = q^{-d_i a_{ij}} F_j$$

$$\sum_{\nu=0}^{1-a_{ij}} (-1)^\nu \begin{bmatrix} 1 - a_{ij} \\ \nu \end{bmatrix}_{q^{2d_i}} E_i^{1-a_{ij}-\nu} E_j E_i^\nu = 0 \quad \text{for all } i \neq j$$

$$\sum_{\nu=0}^{1-a_{ij}} (-1)^\nu \begin{bmatrix} 1 - a_{ij} \\ \nu \end{bmatrix}_{q^{2d_i}} F_i^{1-a_{ij}-\nu} F_j F_i^\nu = 0 \quad \text{for all } i \neq j$$

It is immediate that there is a "decomposition" of $U(\mathfrak{g})$ as

$$U(\mathfrak{g}) \cong \mathbb{C}[F_1, \ldots, F_n] \otimes \mathbb{C}[K_1, \ldots, K_n, K_1^{-1}, \ldots, K_n^{-1}] \otimes \mathbb{C}[E_1, \ldots, E_n].$$

The definition of $U_q(\mathfrak{g})$ is not as bizarre as it may first appear, once one recalls Serre's Theorem that $U(\mathfrak{g})$ is the $\mathbb{C}$-algebra with generators $\{X_i, H_i, Y_i \mid 1 \leq i \leq n\}$ and relations

$$H_i H_j - H_j H_i = 0$$

$$X_i X_j - X_j X_i = \delta_{ij} H_i$$

$$H_i X_j - X_j H_i = a_{ij} X_j \qquad H_i Y_j - Y_j H_i = -a_{ij} Y_j$$

$$(ad X_i)^{1-a_{ij}}(X_j) = 0 \qquad (ad Y_i)^{1-a_{ij}}(Y_j) = 0 \qquad \text{for } i \neq j$$

These last two relations may be rewritten as

$$\sum_{\nu=0}^{1-a_{ij}} (-1)^{\nu} \binom{1-a_{ij}}{\nu} X_i^{1-a_{ij}-\nu} X_j X_i^{\nu} = 0$$

and

$$\sum_{\nu=0}^{1-a_{ij}} (-1)^{\nu} \binom{1-a_{ij}}{\nu} Y_i^{1-a_{ij}-\nu} Y_j Y_i^{\nu} = 0$$

for all $i \neq j$ where $\binom{1-a_{ij}}{\nu}$ is the ordinary binomial coefficient.

The Hopf algebra structure on $U_q(\mathfrak{g})$ is defined by

$$\Delta(E_i) = E_i \otimes K_i^{-1} + K_i \otimes E_i, \quad s(E_i) = -q^{-2d_i} E_i, \quad \varepsilon(E_i) = 0,$$
$$\Delta(F_i) = F_i \otimes K_i^{-1} + K_i \otimes F_i, \quad s(F_i) = -q^{2d_i} F_i, \quad \varepsilon(F_i) = 0,$$
$$\Delta(K_i) = K_i \otimes K_i \qquad\qquad s(K_i) = K_i^{-1}, \qquad \varepsilon(K_i) = 1.$$

Both Δ and ε extend to algebra homomorphisms, and s extends to an algebra anti-homomorphism.

REMARK: If $yx = q^2 xy$, then

$$(x+y)^n = \sum_{j=0}^{n} q^{-j(n-j)} \begin{bmatrix} n \\ j \end{bmatrix}_q x^j y^{n-j}.$$

It is probably better, however, to write this in terms of the Gaussian polynomials. If $n \geq m \geq 0$, define $\begin{Bmatrix} n \\ m \end{Bmatrix} := \frac{(1-t)(1-t^2)\cdots(1-t^{n-m+1})}{(1-t)\cdots(1-t^m)}$ and write $\begin{Bmatrix} n \\ m \end{Bmatrix}_{q^2}$ for its evaluation at $t = q^2$. If $yx = q^2 xy$, then

$$(x+y)^n = \sum_{j=0}^{n} \begin{Bmatrix} n \\ j \end{Bmatrix}_{q^2} x^j y^{n-j}.$$

Representation theory. The representation theory of $U_q(\mathfrak{g})$ is remarkably similar to that of $U(\mathfrak{g})$. This was worked out by Lusztig and Rosso [L1], [R2], [R3]. The classification of finite dimensional simple modules follows the analysis of the classical case, with a few technical complications. Furthermore, every finite dimensional $U_q(\mathfrak{g})$-module is semisimple. The main difference from the classical case is that there are $4^{\operatorname{rank}\mathfrak{g}}$ 1-dimensional $U_q(\mathfrak{g})$-modules, but in a sense this is the only difference.

To classify the finite dimensional simple modules, one begins with $U_q(sl(2))$. However, the ideas for that simple case are the key to the general case. The first observation is that, if V is a finite dimensional $U_q(sl(2))$-module, then both E and F act nilpotently on V. Hence, in a finite dimensional $U_q(\mathfrak{g})$-module there are elements $v_i \neq 0$ such that $E_i.v_i = 0$ for all i. In fact there exists a single $v \neq 0$ such that $E_i.v = 0$ for all i (the analogue of a highest weight vector). To prove this requires a weight theory which we describe next.

First define a partial ordering on the multiplicative group $(\mathbb{C}^*)^n$; thus $(\mathbb{C}^*)^n$ is the analogue of the dual of the Cartan subalgebra. Set $\alpha(j) = (q^{d_1 a_{1j}}, \dots, q^{d_n a_{nj}})$ for $1 \leq j \leq n$. If $\mu, \tau \in (\mathbb{C}^*)^n$, then $\mu \leq \tau \Leftrightarrow \tau = \alpha(1)^{k_1} \alpha(2)^{k_2} \dots \alpha(1)^{k_n} \mu$ for some $(k_1, \dots, k_n) \in \mathbb{N}^n$. Because q is not a root of unity, and because the Cartan matrix is invertible, if $\mu \leq \tau$ and $\tau \leq \mu$ then $\mu = \tau$. Hence this is a partial order.

Let M be a $U_q(\mathfrak{g})$-module. If $\mu = (\mu_1, \dots, \mu_n) \in (\mathbb{C}^*)^n$, define $M_\mu = \{v \in M | K_i.v = \mu_i v \text{ for all } i\}$. If $M_\mu \neq 0$, we call the μ a *weight* of M, and M_μ a *weight space*. It is important to notice that $E_i M_\nu \subset M_\tau$ with $\tau > \nu$, and $F_i M_\mu \subset M_\mu$ with $\mu < \nu$ for all i. A vector $0 \neq v \in M$ is a *highest weight vector* if $v \in M_\nu$ for some ν and $E_i.v = 0$ for all i.

If $v \in M_\lambda$ is a highest weight vector such that $M = U_q(\mathfrak{g}).v$, we call M a *highest weight module*, of *highest weight* λ. For each $\lambda \in (\mathbb{C}^*)^n$, there is a distinguished highest weight module of highest weight λ, namely $M(\lambda) := U_q(\mathfrak{g})/I$ where I is the left ideal generated by the E_i, and the $K_i - \lambda_i$. We call $M(\lambda)$ a *Verma module*. Notice that a highest weight module is a quotient of some Verma module. Since $U_q(\mathfrak{g}) = \mathbb{C}[F_1, \dots, F_n] \oplus I$, $\dim M(\lambda)_\lambda = 1$, and $M(\lambda) = \mathbb{C}[F_1, \dots, F_n].\bar{1}$. A proper submodule of $M(\lambda)$ must be contained in $\sum_{\nu < \lambda} M(\lambda)_\nu$, so there is a unique maximal submodule, and a unique simple quotient module, which we denote by $L(\lambda)$.

Return to V, a finite dimensional simple $U_q(\mathfrak{g})$-module. Since $\dim V < \infty$, there exists a K_i-eigenvector $0 \neq v \in V$. Since $U_q(\mathfrak{g})$ is a sum of K_i-eigenspaces under the action $u \mapsto K_i u K_i^{-1}$, Uv is also a sum of K_i-eigenspaces. Since each K_i is diagonalisable on V, and the K_i commute, the K_i are simultaneously diagonalisable, and $V = \oplus_\mu V_\mu$. Hence V contains a highest weight vector, and is generated by it since V is simple. Therefore $V \cong L(\lambda)$ for some λ. If $\lambda \neq \mu$, then $L(\lambda) \not\cong L(\mu)$. Hence the classification of the finite dimensional simple $U_q(\mathfrak{g})$-modules is reduced to deciding for which λ, $\dim L(\lambda) < \infty$. This is first done for $U_q(sl(2))$, and finally for $U_q(\mathfrak{g})$. The remarkable, and important thing, is that the proof is "identical" to that for $U(\mathfrak{g})$.

THEOREM.

(a) *If $\lambda \in \mathbb{C}^*$ is the highest weight of a finite dimensional simple*

$U_q\big(sl(2)\big)$-module, then $\lambda = \omega q^m$ where $\omega^4 = 1$, and $m \in \{0, 1, \dots\}$.

(b) For every $\omega \in \{\pm 1, \pm i\}$, and every $m \in \mathbb{N}$, ωq^m is the highest weight of an $(m+1)$-dimensional simple $U_q\big(sl(2)\big)$-module. The weights of this module are $\omega q^m, \omega q^{m-2}, \omega q^{m-4}, \dots, \omega q^{-m}$.

THEOREM. *The simple $U_q(\mathfrak{g})$-module of highest weight $\lambda = (\lambda_1, \dots, \lambda_n) \in (\mathbb{C}^*)^n$ is finite dimensional $\Leftrightarrow \lambda_j = \omega_j q^{d_j m_j}$ where $\omega_j \in \{\pm 1, \pm i\}$ and $m_j \in \mathbb{N}$, for all $j = 1, \dots, n$.*

Thus, to a finite dimensional simple $U(\mathfrak{g})$-module there correspond 4^n simple $U_q(\mathfrak{g})$-modules. However, these 4^n modules may all be obtained from a single module by tensoring it with the 4^n 1-dimensional modules. Thus (with an appropriate definition) there are n fundamental representations, and all the representations are obtained by taking the highest weight component of all possible tensor products of the fundamental representations, and then tensoring these with the 1-dimensional representations.

$U_q\big(sl(n)\big)$ **and the Hecke algebra of S_m.** There is a natural representation of $U_q\big(sl(n)\big)$ on $V = \mathbb{C}^n$ defined by $\psi : U_q\big(sl(n)\big) \to \mathrm{End}_{\mathbb{C}} V$ where, for $1 \le i \le n-1$, $E_i \mapsto e_{i,i+1}$, $F_i \mapsto e_{i+1,i}$ and $K_i \mapsto e_{11} + \cdots + e_{i-1,i-1} + q e_{ii} + q^{-1} e_{i+1,i+1} + e_{i+2,i+2} + \cdots + e_{nn}$. Since these elements generate $M_n(\mathbb{C})$, V is a simple module. Notice that, if $q = e^{\frac{1}{4}h}$, then the image of K_i is $\exp\big(\frac{1}{4}h(e_{ii} - e_{i+1,i+1})\big)$; thus this representation of $U_q\big(gl(n)\big)$ is obtained from the standard n-dimensional representation of $gl(n)$ in an obvious way.

By using Δ ($m-1$ times) one obtains a representation of $U_q\big(sl(n)\big)$ on $V^{\otimes m}$ for all m. Define $\varepsilon(j-i) = \begin{cases} +1 & \text{if } j > i \\ -1 & \text{if } j < i \end{cases}$. Define $\widehat{R} : V \otimes V \to V \otimes V$ by $\widehat{R} = q^2(1 \otimes 1) - \sum_{i \ne j}(q^{2\varepsilon(i-j)} e_{ii} \otimes e_{jj} - e_{ij} \otimes e_{ji})$. Thus

$$\widehat{R}(e_i \otimes e_j) = \begin{cases} (q^2 - q^{-2})e_i \otimes e_j + e_j \otimes e_i & \text{if } j > i \\ q^2 e_i \otimes e_i & \text{if } j = i \\ e_j \otimes e_i & \text{if } j < i. \end{cases}$$

Observe that $\widehat{R}$ is a $U_q\big(sl(n)\big)$-module homomorphism, and that the minimal polynomial of $\widehat{R}$ is $(\widehat{R} - q^2)(\widehat{R} + q^{-2}) = 0$. Clearly, $\widehat{R}(e_i \otimes e_j) = e_j \otimes e_i$ for all i, j when $q = 1$.

For each a, $1 \le a \le m-1$, define $T_a : V^{\otimes m} \to V^{\otimes m}$ by

$$T_a = 1 \otimes \cdots \otimes 1 \otimes \widehat{R} \otimes 1 \otimes \cdots \otimes 1$$

where $\widehat{R}$ acts on the ath and $(a+1)^{\text{th}}$ factors in $V^{\otimes m}$. A calculation shows that $T_a T_{a+1} T_a = T_{a+1} T_a T_{a+1}$, and it is clear that $T_a T_b = T_b T_a$ if $|a-b| \ge 2$.

Recall the definition of the *Hecke algebra*, $\mathcal{H}_m(t)$ $(t \in \mathbb{C})$, of the symmetric group S_m. If $s_a = (a, a+1)$ are the usual generators for S_m, then $\mathcal{H}_m$ is the $\mathbb{C}$-vector space with basis s_a and relations $s_a s_{a+1} s_a = s_{a+1} s_a s_{a+1}$, and $s_a s_b = s_b s_a$ if $|a - b| \geq 2$, and $s_a^2 = (t-1)s_a + t$.

If $q = t^4$, then there is a homomorphism $\mathcal{H}_m \to \mathrm{End}_{\mathbb{C}} V^{\otimes m}$ given by $s_a \mapsto q^2 T_a$. Since $\widehat{R}$ is a $U_q(sl(n))$-module map, the image of $\mathcal{H}_m$ actually consists of $U_q(sl(n))$-module maps. The following remarkable result is the analogue of Schur's classical result concerning $GL(n)$ and $\mathbb{C}S_m$. It was also the first way that representations of $U_q(\mathfrak{g})$ were obtained: if $t^m \neq 1$, then $\mathcal{H}_m(t) \cong \mathbb{C}S_m$, and each $\mathcal{H}_m(t)$-isotypic component of $V^{\otimes m}$ is a $U_q(sl(n+1))$-module.

THEOREM. [Jimbo, J3] *Let $U_q(sl(n))$ and $\mathcal{H}_m$ act on $V^{\otimes m}$ as above. Then the images of these algebras in $\mathrm{End}_{\mathbb{C}} V^{\otimes m}$ are mutual commutants of one another, for all except a finite number of $q \in \mathbb{C}$.*

One should therefore think of the Hecke algebra as a quantization of the group algebra of the Weyl group.

Noetherian property and PBW. H. Yamane [Y1], [Y2] has shown that $U_q(sl(n+1))$ has a basis like the Poincaré-Birkoff-Witt basis, and as a consequence $U_q(sl(n+1))$ is a noetherian domain. The basis is constructed from the elements

$$e_{i,i+1} = E_i \quad f_{i,i+1} = F_i$$
$$e_{ij} = q e_{i,j-1} e_{j-1,j} - q^{-1} e_{j-1,j} e_{i,j-1} \quad (j - i > 1)$$
$$f_{ij} = q f_{i,j-1} f_{j-1,j} - q^{-1} f_{j-1,j} f_{i,j} \quad 1 \quad (j - i > 1).$$

Define $(i,j) < (k,m)$ if $i < k$, or if $i = k$ and $j < m$. Yamane shows that, if $q^8 \neq 1$, then a basis for $U_q(sl(n+1))$ is given by

$$f_{k_1,m_1} \cdots f_{k_s,m_s} K_1^{\ell_1} \cdots K_n^{\ell_n} e_{i_1,j_1} \cdots e_{i_t,j_t}$$

with $(k_1, m_1) \leq \cdots \leq (k_s, m_s)$, and $(i_1, j_1) \leq \cdots \leq (i_t, j_t)$ and $(\ell_1, \ldots, \ell_n) \in \mathbb{Z}^n$. With the help of this basis, Yamane defines a filtration such that the associated graded algebra is a localisation of an algebra of the form $\mathbb{C}[y_1, \ldots, y_r]$ defined by relations $y_i y_j = \lambda_{ij} y_j y_i$ for $i \leq j$, where $\lambda_{ij} \in \mathbb{C}$. This is a noetherian domain (see [MP] for a discussion of its properties), and therefore $U_q(sl(n+1))$ is also a noetherian domain. It is not known whether $U_q(\mathfrak{g})$ is a noetherian domain for general $\mathfrak{g}$.

In [L3] Lusztig gives a basis in types A, D and E, and in [L4] improves this to give a basis for $U_q(\mathfrak{g})$ for all $\mathfrak{g}$. The construction depends on defining an action of the braid group of the Weyl group, as algebra automorphisms of $U_q(\mathfrak{g})$.

Duality. The quantum group $\mathcal{O}_q\big(SL(n+1)\big)$ and the q-analogue of the enveloping algebra $U_q\big(sl(n+1)\big)$ are continuous duals to one another, with respect to suitable topologies (this is stated in [FRT]). There is a non-degenerate pairing

$$\langle\,,\,\rangle : \mathcal{O}_q\big(SL(n+1)\big) \times U_q\big(sl(n+1)\big) \to \mathbb{C}$$

satisfying the following (natural) conditions:

(a) The image of the map $\mathcal{O}_q \to U_q^*$ $a \mapsto \langle a,\,\rangle$ belongs to $(U_q)^\circ$, and the image of $U_q \to \mathcal{O}_q^*$ belongs to $(\mathcal{O}_q)^\circ$.

(b) The map $\Psi : \mathcal{O}_q \to \mathrm{End}_{\mathbb{C}}\, U_q$, $\Psi(a)(u) = \sum_{(u)} u_{(1)}\langle a, u_{(2)}\rangle$ makes U_q an $\mathcal{O}_q$-module algebra.

(c) The map $\Phi : U_q \to \mathrm{End}_{\mathbb{C}}\, \mathcal{O}_q$, $\Phi(u)(a) = \sum_{(a)} a_{(1)}\langle u, a_{(2)}\rangle$ makes $\mathcal{O}_q$ a U_q-module algebra.

It turns out that the images of $\mathcal{O}_q$ in U_q° and of U_q in $\mathcal{O}_q^\circ$, although not the full cofinite duals, are the continuous duals with respect to natural cofinite topologies. As so often, the difference from the classical case is due to the 4^n 1-dimensional modules; in the present situation this difference is manifested by the fact that the map $\mathcal{O}_q \to U_q^\circ$ is not surjective. The image of $\mathcal{O}_q$ in U_q° is the full continuous dual with respect to the topology defined by the annihilators of the modules which occur as summands of $V^{\otimes k}$ where V is the standard $(n+1)$-dimensional representation of $U_q\big(sl(n+1)\big)$ i.e., a basis for the topology is the set of ideals $\mathrm{Ann}W$, where $W \subset V^{\otimes k}$.

Skew derivations. Recall that if A is an algebra over the field k, and $\sigma \in \mathrm{Aut}_k A$, then a *skew derivation* or a σ-*derivation* on A is a k-linear map $\delta : A \to A$ such that $\delta(ab) = \delta(a)b + \sigma(a)\delta(b)$ for all $a, b \in A$. To specify the action of a σ-derivation δ it is enough to say how δ acts on generators of the algebra.

Action of $U_q\big(sl(n)\big)$ on $\mathcal{O}_q(\mathbb{C}^n)$. The action of $U_q\big(sl(n)\big)$ on $V = \mathbb{C}^n$ extends in a unique way to an action on the tensor algebra $T(V)$, making $T(V)$ a $U_q\big(sl(n)\big)$-module algebra. The defining ideal of $\mathcal{O}_q(\mathbb{C}^n)$ is a submodule because it is generated by a submodule of $V \otimes V$. Hence $\mathcal{O}_q(\mathbb{C}^n)$ becomes a $U_q\big(sl(n)\big)$-module algebra. Because $\Delta(K_i) = K_i \otimes K_i$, and $K_i^{-1} \in U_q\big(sl(n)\big)$, it follows that each K_i acts as an automorphism of $\mathcal{O}_q(\mathbb{C}^n)$. Moreover, since $\Delta(E_iK_i) = E_iK_i \otimes 1 + K_i^2 \otimes E_iK_i$, and $\Delta(F_iK_i) = F_iK_i \otimes 1 + K_i^2 \otimes F_iK_i$, both E_iK_i and F_iK_i act as skew derivations with respect to the automorphism K_i^2.

The Quantum exterior algebra. Define $\bigwedge_q(\mathbb{C}^n) = \mathbb{C}[\xi_1, \ldots, \xi_n]$ with defining relations $\xi_j\xi_i + q^2\xi_i\xi_j = 0$ for $i \leq j$. A basis is given by $\{\xi_1^{i_1}\cdots\xi_n^{i_n} \mid 0 \leq i_j \leq 1 \text{ for all } j\}$. We may obtain $\bigwedge_q(\mathbb{C}^n)$ in the same way as $\mathcal{O}_q(\mathbb{C}^n)$ *viz.* the $U_q(sl(n))$-submodule of $V \otimes V$ which gives the defining relations of $\mathcal{O}_q(\mathbb{C}^n)$ has a $U_q(sl(n))$-complement which gives the defining relations of $\bigwedge_q(\mathbb{C}^n)$. Furthermore, $\bigwedge_q(\mathbb{C}^n)$ is an $\mathcal{O}_q(M_n(\mathbb{C}))$-comodule via $\lambda(\xi_i) = \Sigma_k X_{ik} \otimes \xi_k$. This extends to an algebra homomorphism $\lambda : \bigwedge_q(\mathbb{C}^n) \to \mathcal{O}_q(M_n(\mathbb{C})) \otimes \bigwedge_q(\mathbb{C}^n)$. Each homogeneous component $\bigwedge_q^m(\mathbb{C}^n)$ is an irreducible $\mathcal{O}_q(M_n(\mathbb{C}))$-comodule. In particular $\bigwedge_q^n(\mathbb{C}^n)$ is a 1-dimensional $\mathcal{O}_q(M_n(\mathbb{C}))$-comodule. If $0 \neq e \in \bigwedge_q^n(\mathbb{C}^n)$, then there is a distinguished element $D \in \mathcal{O}_q(M_n(\mathbb{C}))$ such that $\lambda(e) = D \otimes e$. A computation shows that $D = \det_q X$. Finally, the inclusion $\mathcal{O}_q(M_n(\mathbb{C})) \subset \mathcal{O}_q(GL(n))$ ensures that $\lambda : \bigwedge_q(\mathbb{C}^n) \to \mathcal{O}_q(GL(n)) \otimes \bigwedge_q(\mathbb{C}^n)$ makes $\bigwedge_q(\mathbb{C}^n)$ an $\mathcal{O}_q(GL(n))$-comodule.

Adjoint action. General Hopf algebra considerations give an analogue of the adjoint action of $\mathfrak{g}$ on $U(\mathfrak{g})$. For $u, v \in U_q(\mathfrak{g})$, define $(adu).v = \Sigma_{(u)} u_{(1)}.v.s(u_{(2)})$. To see that this is a module action, first observe that $U_q(\mathfrak{g})$ is a bimodule over itself, and hence a left $U_q(\mathfrak{g}) \otimes U_q(\mathfrak{g})^{op}$-module. Since the antipode s is an algebra anti-homomorphism the following composition is an algebra homomorphism:

$$U_q(\mathfrak{g}) \xrightarrow{\Delta} U_q(\mathfrak{g}) \otimes U_q(\mathfrak{g}) \xrightarrow{id \otimes s} U_q(\mathfrak{g}) \otimes U_q(\mathfrak{g})^{op}.$$

Since this composition defines the adjoint action, the result follows. In contrast to the classical case, this is not a locally finite action; in $U_q(sl(2))$ the repeated adE action on E yields all $E^i K^{i-1}$.

Geometry. There are some obvious questions concerning the geometric aspects of representation theory. Let G be the simply connected, connected semisimple algebraic group with $\text{Lie}G = \mathfrak{g}$, let B be a Borel subgroup, with unipotent radical U.

Are there quantum analogues of the homogeneous spaces (generalised flag varieties) G/P where P is a parabolic subgroup? Taft and Towber [TT] describe a non-commutative algebra which contains subalgebras which should be "the homogeneous coordinate ring of quantum $GL(n)/P$". We shall discuss the case of the Grassmannians.

First recall the classical situation. The exterior algebra $\bigwedge(\mathbb{C}^n)$ admits an action of $GL(n)$ as automorphisms, and each homogeneous component $\bigwedge^n(\mathbb{C}^n)$ is an irreducible representation; these are the fundamental representations. Hence $GL(n)$ acts as automorphisms of the symmetric algebra $S(\bigwedge^m(\mathbb{C}^n))$ on $\bigwedge^m(\mathbb{C}^n)$. The homogeneous coordinate ring of $G_{m,n}$ the Grassmannian of m-dimensional subspaces of $\mathbb{C}^n$ is $S(\bigwedge^m(\mathbb{C}^n))/I$ where I

is an ideal generated by a $GL(n)$-submodule of $S^2(\bigwedge^m(\mathbb{C}^n))$. More precisely, the cone over $G_{m,n}$ is the closed subvariety of $\bigwedge^m(\mathbb{C}^n)$ consisting of all $v_1 \wedge \cdots \wedge v_m$ with $v_j \in \mathbb{C}^n$. The ideal of functions vanishing on this cone is described as follows. Fix a basis $e_1, \ldots, e_n$ for $\mathbb{C}^n$. Let $\mathbb{C}[\wedge(m,n)]$ be the polynomial ring on the set of indeterminates $\wedge(m,n) = \{[\lambda_1, \ldots, \lambda_m] \mid 1 \leq \lambda_1 < \cdots < \lambda_m \leq n$ and $\lambda_i \in \mathbb{N}\}$. It is convenient to write, for each permutation $\sigma \in S_m$, $[\lambda_{\sigma 1}, \ldots, \lambda_{\sigma m}] = \operatorname{sgn}(\sigma)[\lambda_1, \ldots, \lambda_m]$. We view $\mathbb{C}[\wedge(m,n)]$ as the ring of polynomial functions on $\wedge^m(\mathbb{C}^n)$ with $[\lambda_1, \ldots, \lambda_m]$ the dual basis element to $e_{\lambda_1} \wedge \cdots \wedge e_{\lambda_m}$. Then the ideal of functions vanishing on the cone over $G_{m,n}$ is generated by all quadratic Plücker relations

$$\sum_{i=1}^{m+1} (-1)^i [\lambda_1, \ldots, \lambda_{i-1}, \lambda_{i+1}, \cdots, \lambda_{m+1}] [\lambda_i, \mu_1, \ldots, \mu_{m-1}]$$

where $\lambda \in \wedge(m+1, n)$ and $\mu \in \wedge(m-1, n)$. See [DR] and [P].

In the quantum situation the defining relations of the quantum Grassmannian are considerably more complicated—one needs analogues of both the Plücker relations and the commutation relations. These can be found in [TT], and we content ourselves with an explicit description of the "coordinate ring of quantum $G_{2,4}$". Classically this is $\mathbb{C}[X_{12}, X_{13}, X_{14}, X_{23}, X_{24}, X_{34}]$ with the single relation $X_{12}X_{34} - X_{13}X_{24} + X_{14}X_{23} = 0$ where $X_{ij} = [i,j]$ in the above notation.

EXAMPLE: (Quantum $G_{2,4}$). The homogeneous coordinate ring of quantum $G_{2,4}$ is $\mathbb{C}[X_{12}, X_{13}, X_{14}, X_{23}, X_{24}, X_{34}]$ with the following defining relations (the last of which is the analogue of the Plücker relation):

$$X_{cd}X_{ab} = qX_{ab}X_{cd} \qquad \text{if } \{a, b, cd\} \neq \{1, 2, 3, 4\} \text{ and } ab < cd \text{ where}$$
we use the lexicographical ordering $12 < 13 < 14 < 23 < 24 < 34$

$$X_{23}X_{14} = X_{14}X_{23}$$
$$X_{24}X_{13} = X_{13}X_{24} + (q - q^{-1})X_{14}X_{23}$$
$$X_{34}X_{12} = X_{12}X_{34} + (q - q^{-1})X_{13}X_{24} + (q^{-2} - 1)X_{14}X_{23}$$
$$qX_{12}X_{34} - X_{13}X_{24} + q^{-1}X_{14}X_{23} - 0.$$

The action of $U(\mathfrak{g})$ as differential operators on $\mathcal{O}(G/U)$ is such that each finite dimensional simple $\mathfrak{g}$-module appears in $\mathcal{O}(G/U)$ with multiplicity 1. What is the analogue of $\mathcal{O}(G/U)$ for $U_q(\mathfrak{g})$? The theorem of Borel-Weil-Bott says that every finite dimensional irreducible representation of $\mathfrak{g}$ may be realised as $H^0(G/B, \mathcal{L})$ for a suitable line bundle $\mathcal{L}$ on G/B. Is there a quantum version of Borel-Weil-Bott? What is the quantum version of $\mathcal{O}_{G/B}$, and what are the line bundles?

In [PW1], analogues of the cohomology functors $H^i(G/B, -)$ are constructed. These are defined as the right derived functors of the induction functor from the quantised Borel subgroup for quantum $SL(n)$. This is a purely algebraic construction, and is done just for $SL(n)$ and $GL(n)$. In particular, there is no geometric interpretation of a "quantum flag variety". Nevertheless, it is shown in [PW2] that these cohomology groups are very well-behaved: it is a challenging problem to extend the ideas in [PW1] and [PW2] to other groups than $SL(n)$ and $GL(n)$. The ideas and results in [PW1] and [PW2] concerning analogues of $H^i(G/B, -)$ are also given in [An].

§4. The other groups, and R-matrices

Although $U_q(\mathfrak{g})$ has been defined for all $\mathfrak{g}$, only $\mathcal{O}_q(SL(n))$ and $\mathcal{O}_q(GL(n))$ have been defined on the dual side. Are there versions of $\mathcal{O}_q(SO(n))$, $\mathcal{O}_q(SP(2n))$ etc.? To explain the answer, which is "yes", we adopt the approach introduced in [FRT]. The advantage of this approach is that the rather complicated defining relations can be expressed succinctly, and more conceptually. More detailed information about quantum $SO(n)$ and $SP(2n)$ is given in the papers of Takeuchi [T1], [T2]. The presentation there differs slightly from ours. In particular, he works with the matrix $\widehat{R}$ rather than with R (see below). In particular, he says that the Diamond Lemma may be used to obtain a PBW basis for the quantum symplectic, and quantum Euclidean spaces described below.

The coordinate ring of quantum $n \times n$ matrices. Let $R \in M_n(\mathbb{C}) \otimes M_n(\mathbb{C}) = M_{n^2}(\mathbb{C})$. Let $e_1, \ldots, e_n$ be a basis for $\mathbb{C}^n$, and $e_i \otimes e_j$ a basis for $\mathbb{C}^n \otimes \mathbb{C}^n$ on which R acts from the left. Let $\mathbb{C}\langle t_{ij} \rangle$ be the free algebra on indeterminates t_{ij}, $1 \leq i, j \leq n$, and define I_R to be the two-sided ideal generated by the entries in the $n^2 \times n^2$ matrix

$$RT_1T_2 - T_2T_1R$$

where $T_1 = T \otimes I, T_2 = I \otimes T, T = (t_{ij}) \in M_n(\mathbb{C}\langle t_i \rangle)$, and I is the $n \times n$ identity matrix. Define the coordinate ring of the *quantum matrix algebra* (associated to R) to be

$$\mathcal{O}_R(M_n(\mathbb{C})) := \mathbb{C}\langle t_{ij} \rangle / I_R.$$

It is quite easy to see that if $R = I \otimes I$ is the identity matrix, then $\mathcal{O}_R(M_n(\mathbb{C})) \cong \mathcal{O}(M_n(\mathbb{C}))$. There is a bialgebra structure on $\mathcal{O}_R(M_n(\mathbb{C}))$ given by $\Delta(t_{ij}) = \sum_k t_{ik} \otimes t_{kj}$, and $\varepsilon(t_{ij}) = \delta_{ij}$.

REMARK: View a tensor product of two $n \times n$-matrices as an $n \times n$ matrix of blocks, each block being an $n \times n$ matrix. If R is an $n^2 \times n^2$-matrix write $R_{ijk\ell}$ for the coefficient of $e_i \otimes e_j$ in $R(e_k \otimes e_\ell)$. In some of the Russian papers our $R_{ijk\ell}$ is written as $R_{ij}^{k\ell}$. If we order the basis $e_1 \otimes e_1, e_1 \otimes e_2, \dots, e_1 \otimes e_n, e_2 \otimes e_1, \dots, e_n \otimes e_{n-1}, e_n \otimes e_n$ then $R_{ijk\ell}$ is the entry in the $j\ell^{\text{th}}$ position of the ik^{th} block (that is, in the row labelled ij, and the column labelled $k\ell$). If e_{ij} are the usual matrix units for $M_n(\mathbb{C})$ with respect to the basis $e_1, \dots, e_n$, then $e_{ij} \otimes e_{k\ell} = e_{ikj\ell}$, the matrix with 1 in row ik and column $j\ell$.

If A and B are $n \times n$ matrices, then $A \otimes B$ is the $n^2 \times n^2$ matrix with $a_{ij}b_{k\ell}$ as the $k\ell^{\text{th}}$ entry in the ij^{th} block. Thus T_1 has as its $k\ell^{\text{th}}$ block $\text{diag}(t_{k\ell}, \dots, t_{k\ell})$, and T_2 has in its kk^{th} block a copy of T, and the off diagonal blocks are zero. Therefore the defining equations of I_R are $R_{ijk\ell}t_{km}t_{\ell p} = t_{j\ell}t_{ik}R_{k\ell m p}$ where we adopt the convention of summation over repeated indices; hence this is a sum over k and ℓ, which holds for every 4-tuple (i, j, m, p).

EXAMPLE: 1. Set

$$T = \begin{bmatrix} a & b \\ c & d \end{bmatrix} \qquad R = \begin{bmatrix} q^2 & 0 & 0 & 0 \\ 0 & 1 & 0 & 0 \\ 0 & p & 1 & 0 \\ 0 & 0 & 0 & q^2 \end{bmatrix} \qquad \text{where } p = q^2 - q^{-2}.$$

This gives 16 relations, which reduce to the 6 defining relations for $\mathcal{O}_q(M_2(\mathbb{C}))$. Thus $\mathcal{O}_R(M_n(\mathbb{C})) \cong \mathcal{O}_q(M_2(\mathbb{C}))$.

The coordinate ring of quantum n-space. Let $f \in \mathbb{C}[y]$. Define $\sigma : \mathbb{C}^n \otimes \mathbb{C}^n \to \mathbb{C}^n \otimes \mathbb{C}^n$ by $\sigma(u \otimes v) = v \otimes u$, and set $\widehat{R} = \sigma R$; thus $\widehat{R}_{ijk\ell} = R_{jik\ell}$. Let $I_{f,R}$ be the ideal in the free algebra $\mathbb{C}\langle X_1, \dots, X_n \rangle$ generated by the entries in $f(\widehat{R})\widetilde{X}$ where $\widetilde{X}$ is the $n^2 \times 1$ column matrix with $X_i X_j$ in the row labelled by ij. Hence the generators of $I_{f,R}$ are (with the usual summation convention)

$$f(\widehat{R})_{ijk\ell} X_k X_\ell = 0 \quad \text{for all } i, j.$$

Define the coordinate ring of the *quantum n-space* associated to f and R to be

$$\mathcal{O}_{f,R}(\mathbb{C}^n) := \mathbb{C}\langle X_1, \dots, X_n \rangle / I_{f,R}.$$

There is an algebra homomorphism $\lambda : \mathcal{O}_{f,R}(\mathbb{C}^n) \rightarrow \mathcal{O}_R(M_n(\mathbb{C})) \otimes \mathcal{O}_{f,R}(\mathbb{C}^n)$ defined by $\lambda(X_i) = \sum_k t_{ik} \otimes X_k$. This gives $\mathcal{O}_{f,R}(\mathbb{C}^n)$ the structure of a left comodule for $\mathcal{O}_R(M_n(\mathbb{C}^n))$. Thus we say that the quantum matrices act on the quantum n-space.

EXAMPLE: 2. If R is as in the previous example, take $f(y) = y - q^2$. Then $\mathcal{O}_{f,R}(\mathbb{C}^2) = \mathbb{C}[X_1, X_2]$ with $X_2 X_1 = q^{-2} X_1 X_2$. That is $\mathcal{O}_{f,R}(\mathbb{C}^2) \cong \mathbb{C}_q[X, Y]$, the quantum plane as defined in §2.

Now, with the same R, take $f(y) = y + q^{-2}$. In this case $\mathcal{O}_{f,R}(\mathbb{C}^2) = \mathbb{C}[X_1, X_2]$ with $X_1^2 = X_2^2 = X_2 X_1 + q^2 X_1 X_2 = 0$, and $\mathcal{O}_{f,R}(\mathbb{C}^2) \cong \Lambda_q(\mathbb{C}^2)$, the quantum exterior algebra as defined in §2.

The comodule maps agree with those in §2.

Notice that these two choices of f come from the minimal polynomial of $\widehat{R}$, namely $(\widehat{R} - q^2)(\widehat{R} + q^{-2})$.

$\mathcal{O}_q(M_n(\mathbb{C}))$ **and** $\mathcal{O}_q(\mathbb{C}^n)$. The quantum matrices, and quantum n-space defined in §3 may also be obtained from this general construction. Take

$$R = q^2 \sum_{i,j} e_{ij} \otimes e_{ji} - \sum_{i \neq j} (q^{2\varepsilon(i-j)} e_{ji} \otimes e_{ij} - e_{ii} \otimes e_{jj})$$

$$= q^2 \sum_i e_{ii} \otimes e_{ii} + \sum_{i \neq j} e_{ii} \otimes e_{jj} + (q^2 - q^{-2}) \sum_{i>j} e_{ij} \otimes e_{ji}$$

$$= q^2 \sum_i e_{iiii} + \sum_{i \neq j} e_{ijij} + (q^2 - q^{-2}) \sum_{i>j} e_{ijji}.$$

Then $\mathcal{O}_R(M_n(\mathbb{C})) \cong \mathcal{O}_q(M_n(\mathbb{C}))$, and (with $f(y) = y - q^2$), $\mathcal{O}_{f,R}(\mathbb{C}^n) \cong \mathcal{O}_q(\mathbb{C}^n)$. An explicit description of the action of $\widehat{R}$ on the basis $e_i \otimes e_j$ already appeared in the discussion of the Hecke algebra as $\mathrm{End}_{U_q(sl(n))}(\mathbb{C}^n \otimes \mathbb{C}^n)$, in §3. In particular, $\widehat{R} : \mathbb{C}^n \otimes \mathbb{C}^n \rightarrow \mathbb{C}^n \otimes \mathbb{C}^n$ is a $U_q(sl(n))$-module map (with the earlier module action). Hence $f(\widehat{R})$ is a $U_q(sl(n))$-module map, and thus $\ker f(\widehat{R})$ is a $U_q(sl(n))$-submodule of $\mathbb{C}^n \otimes \mathbb{C}^n$. Hence the quotient algebra of the tensor algebra $T(\mathbb{C}^n)/\langle \ker f(\widehat{R}) \rangle$ can be made into a $U_q(sl(n))$-module algebra in a natural way.

Define the *quantum exterior algebra* $\Lambda_q(\mathbb{C}^n) := \mathcal{O}_{g,R}(\mathbb{C}^n)$ where $g(y) = y + q^{-2}$. The previous considerations show that $\Lambda_q(\mathbb{C}^n)$ is a comodule for $\mathcal{O}_q(\mathbb{C}^n)$. Clearly $\Lambda_q(\mathbb{C}^n)$ is graded $\Lambda_q(\mathbb{C}^n) = \oplus \Lambda_q(\mathbb{C}^n)_j$. It is also finite dimensional, and its top degree component is of degree n. Moreover, if $\xi_1, \ldots, \xi_n$ is a basis for $\mathbb{C}^n = \Lambda_q(\mathbb{C}^n)_1$, then $\xi_1 \ldots \xi_n$ spans $\Lambda_q(\mathbb{C}^n)_n$. Hence there is a distinguished element, D say, in $\mathcal{O}_q(M_n(\mathbb{C}^n))$ such that $\lambda(\xi_1 \ldots \xi_n) = D \otimes \xi_1 \ldots \xi_n$. In fact, $D = \det_q X$ as defined in §3. This gives a natural interpretation/definition of the quantum determinant which nicely extends the classical definition.

Quantum $SP(2n)$. We will define $\mathcal{O}_q\big(SP(2n)\big)$ as a certain quotient ring of $\mathcal{O}_R\big(M_{2n}(\mathbb{C})\big)$ but for a *different* R than that above. Set

$$R = q\sum_{i=1}^{2n} e_{ii} \otimes e_{ii} + \sum_{i \neq j,j'} e_{ii} \otimes e_{jj} + q^{-1}\sum_{i=1}^{2n} e_{ii} \otimes e_{i'i'}$$

$$+ (q - q^{-1})\sum_{i>j} e_{ij} \otimes e_{ji} - (q - q^{-1})\sum_{i>j} q^{\rho(i)-\rho(j)}\varepsilon_i\varepsilon_j e_{ij} \otimes e_{i'j'}$$

where $i' = (2n+1) - i$, $\big(\rho(1),\dots,\rho(2n)\big) = (n, n-1, \dots, 1, -1, \dots, -n)$, $\varepsilon_i = 1$ if $1 \leq i \leq n$, and $\varepsilon_i = -1$ if $n+1 \leq i \leq 2n$. Define

$$C = \text{anti-diag}(q^n, q^{n-1}, \dots, q, 1, -1, -q^{-1}, \dots, -q^{-n}) \in M_{2n}(\mathbb{C}).$$

Define

$$\mathcal{O}_q\big(SP(2n)\big) := \mathcal{O}_R\big(M_{2n}(\mathbb{C})\big)/\langle TCT^tC^{-1} - I, CT^tC^{-1}T - I\rangle$$

where I is the $2n \times 2n$ identity matrix, $T = (t_{ij})$, and T^t is the transpose. If we set $Y = TC = (y_{ij})$, then $\mathcal{O}_q\big(SP(2n)\big) = \mathcal{O}_R\big(M_{2n}(\mathbb{C})\big)/\langle YY^t = -I = Y^tY\rangle$.

The coordinate ring of *quantum symplectic 2n-dimensional space* is defined as $\mathcal{O}_q(sp\mathbb{C}^{2n}) := \mathcal{O}_{f,R}(\mathbb{C}^{2n})$ where $f = y - q$, and R is the R just specified, *not* the R used in the construction of $\mathcal{O}_q(\mathbb{C}^n)$. Therefore $\mathcal{O}_q(sp\mathbb{C}^{2n}) = \mathbb{C}[X_1, \dots, X_{2n}]$ with relations

$$X_iX_j = qX_jX_i \qquad \text{for } 1 \leq i < j \leq 2n \text{ and } j \neq i',$$

$$X_{i'}X_i = X_iX_{i'} + (q^2 - 1)\sum_{j=1}^{i'-1} q^{\rho(i')-\rho(j)}\varepsilon_{i'}\varepsilon_j X_jX_{j'} \qquad \text{for } 1 \leq i < i' \leq 2n$$

EXAMPLE: 3. $(n = 2)$ Then $\mathcal{O}_q(sp\mathbb{C}^4) = \mathbb{C}[X_1, X_2, X_3, X_4]$ with $X_1X_2 = qX_2X_1, X_1X_3 = qX_3X_1, X_2X_4 = qX_4X_2, X_3X_4 = qX_4X_3, X_2X_3 = q^2X_3X_2 + (q - q^{-1})X_1X_4, X_4X_1 = q^{-2}X_1X_4$. This is an iterated Ore extension. What are its (homogeneous) prime and primitive ideals?

Quantum $SO(n)$. We will define $\mathcal{O}_q\big(SO(n)\big)$ as a certain quotient ring of $\mathcal{O}_R\big(M_n(\mathbb{C})\big)$ but (again) for a *different* R. If n is even, set

$$R = q\sum_{i \neq i'} e_{ii} \otimes e_{ii} + \sum_{i \neq j,j'} e_{ii} \otimes e_{jj} + q^{-1}\sum_{i \neq i'} e_{ii} \otimes e_{i'i'}$$

$$+ (q - q^{-1})\sum_{i>j} e_{ij} \otimes e_{ji} - (q - q^{-1})\sum_{i>j} q^{\rho(i)-\rho(j)} e_{ij} \otimes e_{i'j'}$$

where $i' = (n+1) - i$, and $\rho = \big(\rho(1), \ldots, \rho(n)\big) = (\frac{1}{2}n - 1, \ldots, 1, 0, 0,$
$-1, \ldots, 1 - \frac{1}{2}n)$. If $n = 2k + 1$, then R is as above, except that one adds
$e_{kk} \otimes e_{kk}$, and $\big(\rho(1), \ldots, \rho(n)\big) = (\frac{1}{2}n - 1, \ldots, \frac{1}{2}, 0, -\frac{1}{2}, \ldots, 1 - \frac{1}{2}n)$. Set

$$C = q^\rho .\text{anti-diag}(1, 1, \ldots, 1) \in M_n(\mathbb{C}).$$

Define

$$\mathcal{O}_q\big(SO(n)\big) := \mathcal{O}_R\big(M_n(\mathbb{C})\big)/\langle TCT^tC^{-1} - I, CT^tC^{-1}T - I \rangle.$$

If we write $Y = TC$, and consider $Y = (y_{ij})$ as a matrix of indeterminates
generating $\mathcal{O}_q\big(SO(n)\big)$, then

$$\mathcal{O}_q\big(SO(n)\big) = \mathcal{O}_R\big(M_n(\mathbb{C})\big)/\langle YY^t = I = Y^tY \rangle.$$

The coordinate ring of *quantum Euclidean n-dimensional space* is defined
as $\mathcal{O}_q(so\mathbb{C}^n) := \mathcal{O}_{f,R}(\mathbb{C}^n)$ where $f = y^2 - (q + q^{1-n})y + q^{2-n}$, and R is as
above. Therefore $\mathcal{O}_q(so\mathbb{C}^n) = \mathbb{C}[X_1, \ldots, X_n]$ with relations $X_iX_j = qX_jX_i$
for $1 \leq i < j \leq n$ and $j \neq i'$, and

$$X_{i'}X_i = X_iX_{i'} + (q^n - q^{n-2})(1 + q^{n-2})^{-1}\sum_{j=1}^{i'-1} q^{\rho(i')-\rho(j)}X_jX_{j'} +$$

$$- (q^2 - 1)(1 + q^{n-2})^{-1}\sum_{j=i'}^{n} q^{\rho(i')-\rho(j)}X_jX_{j'} \qquad \text{for } 1 \leq i < i' \leq n$$

EXAMPLE: 4. ($n = 3$) Then $\mathcal{O}_q(so\mathbb{C}^3) = \mathbb{C}[X_1, X_2, X_3]$ with $X_1X_2 = qX_2X_1$, $X_2X_3 = qX_3X_2$, $X_1X_3 - X_3X_1 = (q^{\frac{1}{2}} - q^{-\frac{1}{2}})X_2^2$. This is an
iterated Ore extension. What are its (homogeneous) prime and primitive
ideals? In fact, this is a 3-dimensional regular algebra appearing in [AS],
and the cubic divisor in $\mathbb{P}^2$ associated to it (à la [ATV]) is the union of a
line and a conic.

REMARK: The choices of f in defining $\mathcal{O}_q(sp\mathbb{C}^{2n})$ and $\mathcal{O}_q(so\mathbb{C}^n)$ are motivated by the minimal polynomial satisfied by $\widehat{R}$. For the symplectic group
the minimal polynomial is $(\widehat{R} - q)(\widehat{R} + q^{-1})(\widehat{R} + q^{1-n})$, and for the orthogonal group it is $(\widehat{R} - q)(\widehat{R} + q^{-1})(\widehat{R} - q^{1-n})$ (see e.g. [T1]).

These matrices R have their origin in the Yang-Baxter equation.

The Yang-Baxter equation. Quantum groups arose from the Yang-
Baxter equation which appears in statistical mechanics. This author is not

competent to explain the physical origin and meaning of the equation, so
only the algebraic aspects will be discussed.

Let V be a finite dimensional vector space and $R \in \mathrm{End}_{\mathbb{C}}(V \otimes V)$. If $R = \sum_i a_i \otimes b_i$, define $R_{12}, R_{13}, R_{23} \in \mathrm{End}(V \otimes V \otimes V)$ by $R_{12} = \sum_i a_i \otimes b_i \otimes 1$, $R_{13} = \sum_i a_i \otimes 1 \otimes b_i$, $R_{23} = \sum_i 1 \otimes a_i \otimes b_i$; that is R_{ij} acts via R on the ith and j^{th} copy of V. The (constant) *quantum Yang-Baxter equation* (QYBE) is

$$R_{12} R_{13} R_{23} = R_{23} R_{13} R_{12}.$$

Trivial solutions are $R = I$, where I is the identity, and $R = \sigma$ where $\sigma(u \otimes v) = v \otimes u$. Suppose that R is a solution depending differentiably on a parameter $h \in \mathbb{C}$, $R = I + r_1 h + r_2 h^2 + \cdots$. Then $r := r_1 \in \mathrm{End}(V \otimes V)$ must satisfy the equation

$$[r_{12}, r_{13}] + [r_{12}, r_{23}] + [r_{13}, r_{23}] = 0.$$

This is the (constant) *classical Yang-Baxter equation* (CYBE). Here r_{ij} has the same meaning as R_{ij}.

For physical reasons it is more natural to consider the QYBE equation for R depending on parameters u, v

$$R_{12}(u) R_{13}(u+v) R_{23}(v) = R_{23}(v) R_{13}(u+v) R_{12}(u).$$

In the limit this gives the equation

$$[r^{12}(u), r^{13}(u+v)] + [r^{12}(u), r^{23}(v)] + [r^{13}(u+v), r^{23}(v)] = 0.$$

Since the CYBE just involves commutators it is natural to look for a solution $r(u) \in \mathfrak{g} \otimes \mathfrak{g}$ where $\mathfrak{g}$ is a Lie algebra. Given such a "universal" solution, then each $\mathfrak{g}$-module V gives a "particular" solution to the CYBE. Hence one may proceed to find all solutions inside $\mathfrak{g} \otimes \mathfrak{g}$, and then find all representations of $\mathfrak{g}$ to find solutions to CYBE. For a certain class of solutions this was done for all simple $\mathfrak{g}$ by Belavin and Drinfeld [BD]. For example, one solution is $r(u) = \Omega/u$ where $\Omega \in \mathfrak{g} \otimes \mathfrak{g}$ is the Casimir element; that is, $[\Omega, 1 \otimes X + X \otimes 1] = 0$ for all $X \in \mathfrak{g}$.

Given a solution to the CYBE in $\mathfrak{g} \otimes \mathfrak{g}$, Kulish, Reshetikhin and Sklyanin [KRS] proposed to quantize these solutions: is there a solution to QYBE (depending on a parameter h) such that the limit (as $h \to 0$) is the given solution to CYBE? It was realised that to do this one needed to "quantize" $U(\mathfrak{g})$. Motivated by Kulish and Reshetikhin [KR] who quantized $U(sl(2))$, Drinfeld [D1] and Jimbo [J1] independently obtained quantizations, constructing algebras depending on a parameter h, such that the limit (as

$h \to 0$) was the original $U(\mathfrak{g})$. One such algebra is $U_h(\mathfrak{g})$ as described for $\mathfrak{g} = sl(2)$ in Section 2. We have not discussed another quanitization obtained by Drinfeld. He calls it the Yangian of $\mathfrak{g}$, and denotes it by $Y(\mathfrak{g})$; see [D2], [D3] and [D5] for details.

The matrices R which were used in the construction of quantum $SL(n)$, $SP(2n)$ and $SO(n)$ are all solutions to the QYBE. There are other solutions to the QYBE which we have not discussed. These lead to other non-commutative algebras; see [S1] and [S2].

REMARK: Suppose that $R \in \mathrm{End}(\mathbb{C}^n \otimes \mathbb{C}^n)$ satisfies the QYBE. Then there is a distinguished n-dimensional $\mathcal{O}_R(M_n(\mathbb{C}))$-module, given by sending t_{ij} to the $n \times n$ matrix with $k\ell^{\mathrm{th}}$ entry $R_{ikj\ell}$. Taking Example 1 above, this gives the 2-dimensional $\mathcal{O}_q(M_2(\mathbb{C}))$-module defined by

$$a \mapsto \begin{bmatrix} q^2 & 0 \\ 0 & 1 \end{bmatrix} \qquad b \mapsto 0 \qquad c \mapsto \begin{bmatrix} 0 & p \\ 0 & 0 \end{bmatrix} \qquad d \mapsto \begin{bmatrix} 1 & 0 \\ 0 & q^2 \end{bmatrix}.$$

Yang-Baxter, braids and knots. Recently V. F. R. Jones constructed a new invariant for knots, now called the Jones polynomial ([Jo1], [Jo2]). These polynomials were constructed through a trace on the hyperfinite type II_1 factor, A say. This is the unique von Neumann algebra with center $\mathbb{C}$ which is the closure of a union of finite dimensional subalgebras. In particular, A can be constructed as the closure of a direct limit of the Hecke algebras, $\mathcal{H}_m(t)$ for the symmetric group S_m (here $t \in \mathbb{C}$ is a parameter). Since each $\mathcal{H}_m(t)$ is a quotient of the group algebra of the infinite braid group, this gives a map from the braid group $B_\infty \to A$. Taking traces therefore gives a $\mathbb{C}$-valued function on B_∞. A knot may be represented by a closed braid. Two braids will give the same knot if and only if one can be obtained from the other by a sequence of Markov moves. Since the trace is invariant with respect to Markov moves, one obtains a $\mathbb{C}$-valued function on knots. This function depends on the parameter t, so a fixed knot determines a function $\mathbb{C} \to \mathbb{C}$. After a suitable adjustment, this function can be made into a polynomial, the Jones polynomial.

We now explain how solutions to the Yang-Baxter equation determine representations of the braid group. Recall that the braid group, B_n say, is generated by elements $\beta_i (1 \le i \le n)$ with relations $\beta_i \beta_j = \beta_j \beta_i$ if $|j-i| \ge 2$, and $\beta_i \beta_{i+1} \beta_i = \beta_{i+1} \beta_i \beta_{i+1}$. Let $R \in \mathrm{End}(V \otimes V)$, and write $\widehat{R} = \sigma R$ where $\sigma : V \otimes V \to V \otimes V$ is given by $\sigma(x \otimes y) = y \otimes x$. Then R satisfies the QYBE if and only if $\widehat{R}_{12} \widehat{R}_{23} \widehat{R}_{12} = \widehat{R}_{23} \widehat{R}_{12} \widehat{R}_{23}$. Hence there is a representation of B_n on $V^{\otimes n}$ defined by

$$\beta_i.(v_1 \otimes v_2 \otimes \cdots \otimes v_n) = v_1 \otimes \cdots \otimes v_{i-1} \otimes \widehat{R}(v_i \otimes v_{i+1}) \otimes v_{i+2} \otimes \cdots \otimes v_n.$$

If $\widehat{R}$ also satisfies $\widehat{R}^2 = 1$, then this representation of B_n factors through a representation of S_n.

Connections between the Yang-Baxter equation, representations of quantum groups and invariants of knots are discussed in [Jo3], [Ko], [KR1], [Re2], [RT], [Tu].

§5. Quadratic algebras

A *quadratic algebra* is an algebra $A = T(V)/I$, where V is a finite dimensional vector space, and I is an ideal generated by a subspace of $V \otimes V$. Write A_n for the image of the nth tensor power $T^n(V)$. Thus $A = \bigoplus_n A_n$ is $\mathbb{N}$-graded, and generated by A_1. Notice that the quantum affine, symplectic, Euclidean spaces, and $\mathcal{O}_R(M_n(\mathbb{C}))$, are all quadratic algebras.

The ubiquity of quadratic algebras is indicated by the following result [BF]. Let A be a finitely presented graded algebra which is generated in degree 1. Define $A(d)$ to be the subalgebra generated by the homogeneous elements of degree d. Then, for sufficiently large d, $A(d)$ is defined by quadratic relations.

It is worth noticing that $A(d)$ is the ring of invariants for the action of the cyclic group $\mathbb{Z}/d\mathbb{Z}$ on A, where the generator $\omega = e^{2\pi i/d}$ acts on A_n as scalar multiplication by ω^n. General results on group actions show that there is a close relationship between the properties of A and those of $A(d)$.

If A is commutative, then $\text{Proj}A(d) = \text{Proj}A$, so from the point of view of projective algebraic geometry, there is no loss of generality in only considering algebras defined by quadratic relations. That is given X, there exists an embedding $X \to \mathbb{P}^n$ such that the defining ideal of X is generated by quadratics.

In addition to the preceeding examples, there are a number of other significant algebras defined by quadratic relations (see [BG]). Manin's influential Montréal notes [M2] on quantum spaces are set in the context of algebras defined by quadratic relations.

EXAMPLE: An enveloping algebra $U(\mathfrak{g})$ can be written as a quotient of a quadratic algebra in a canonical way. If $X_1, \dots, X_n$ is a basis of $\mathfrak{g}$, then write $\widetilde{U}(\mathfrak{g}) = \mathbb{C}[X_1, \dots, X_n, t]$ defined by the relations $X_i t = t X_i$ for all i, and $X_i X_j - X_j X_i = t[X_i, X_j]$ for all i, j. Notice that $U(\mathfrak{g}) \cong \widetilde{U}(\mathfrak{g})/\langle t - 1 \rangle$. It might be interesting to know more about the structure of $\widetilde{U}(\mathfrak{g})$; for example, what is its primitive spectrum, and in particular how does the primitive spectrum of $U(\mathfrak{g}_\lambda) := \widetilde{U}(\mathfrak{g})/\langle t - \lambda \rangle$ vary with $\lambda \in \mathbb{C}$.

Given a quadratic algebra $A = T(V)/\langle W \rangle$ where $W \subset V \otimes V$, define the *dual* of A as the quadratic algebra $A^! := T(V^*)/\langle W^\perp \rangle$ where $W^\perp \subset V^* \otimes V^*$

is the orthogonal to W. For example, if A is the symmetric algebra on V, then $A^!$ is the exterior algebra on V^*. More interesting for us is that $\mathcal{O}_q(\mathbb{C}^n)$ and $\Lambda_q(\mathbb{C}^n)$ are duals to one another.

Manin's construction of quantum matrix spaces. We now describe a general procedure due to Manin ([M1] and [M2]), which constructs from a given quadratic algebra, a new quadratic algebra. The new quadratic algebra is a bialgebra, and has both the original algebra and its dual as comodules. Applying this procedure to $\mathcal{O}(\mathbb{C}^n)$ yields $\mathcal{O}(M_n(\mathbb{C}))$, and applying the procedure to $\mathcal{O}_q(\mathbb{C}^n)$ yields $\mathcal{O}_q(M_n)$.

Let $A = T(V)/\langle W \rangle$ where $W \subset V \otimes V$. Define

$$\underline{\mathrm{end}}(A) = T(V^* \otimes V)/\langle s_{23}(W^\perp \otimes W) \rangle$$

where s_{23} swaps the second and third tensor entries. Fix a basis $x_1, \dots, x_n$ for V, and let $\xi_1, \dots, \xi_n$ be the dual basis in V^*. Write $z_{ij} = \xi_i \otimes x_j$. There is a bialgebra structure on $\underline{\mathrm{end}}(A)$ defined by $\Delta(z_{ij}) = \sum_k z_{ik} \otimes z_{kj}$, and $\varepsilon(z_{ij}) = \delta_{ij}$. There is an algebra homomorphism $\lambda : A \to \underline{\mathrm{end}}(A) \otimes A$ defined by $\lambda(x_i) = \sum_k z_{ik} \otimes x_k$, which makes A an $\underline{\mathrm{end}}(A)$ comodule. Unfortunately, $\underline{\mathrm{end}}(A)$ is not the algebra we want; it is too large, as the next example shows.

EXAMPLE: Consider $A = \mathcal{O}_q(\mathbb{C}^2) = \mathbb{C}_q[X, Y]$, where $XY - q^2 YX = 0$. Let $\{\xi, \eta\}$ be the dual basis to $\{X, Y\}$. Then $A^! = \mathbb{C}[\xi, \eta]$ with $\xi^2 = \eta^2 = \xi\eta + q^{-2}\eta\xi = 0$. Set $a = \xi \otimes X = z_{11}$, $b = \xi \otimes Y = z_{12}$, $c = \eta \otimes X = z_{21}$, $d = \eta \otimes Y = z_{22}$. Thus the defining relations for $\underline{\mathrm{end}}(A)$ are

$$ab - q^2 ba = 0, \quad cd - q^2 dc = 0, \quad ad - q^2 bc + q^{-2} cb - da = 0.$$

These are half of the relations required to define $\mathcal{O}_q(M_2(\mathbb{C}))$. If b and c are swapped in the above three relations, then the resulting 6 relations define $\mathcal{O}_q(M_2(\mathbb{C}))$.

The previous example is the basis for the general construction of Manin. Let $t : V^* \otimes V \to V^* \otimes V$ be the transpose $t(z_{ij}) = z_{ji}$. Extend t to an automorphism of the free algebra $\mathbb{C}\langle z_{ij} \rangle$. Define

$$\underline{\mathrm{e}}(A) = T(V^* \otimes V)/\langle S_{23}(W^\perp \otimes W) + t S_{23}(W^\perp \otimes W) \rangle.$$

Call $\underline{\mathrm{e}}(A)$ the *quantum matrix algebra* associated to A. In fact, $\underline{\mathrm{e}}(A)$ depends on the choice of basis for V, just as the transpose of a matrix depends on a choice of basis, or on the choice of a non-degenerate symmetric bilinear form on V. Hence we should first choose such a form, g say, and write $\underline{\mathrm{e}}(A, g)$. Some examples are given in [M1] and [M2].

The next step is to obtain a Hopf algebra from $\underline{e}(A)$. If we were to try defining an antipode, s say, on $\underline{e}(A)$, then the Hopf algebra axioms would force $Zs(Z) = I = s(Z)Z$, where $Z = (z_{ij} \in M_n(\underline{e}(A)))$, and I is the $n \times n$ identity matrix. Hence if we formally invert $s(Z)$ à la P. M. Cohn (i.e., freely adjoin elements to $\underline{e}(A)$, and factor out by the required relations) then one obtains a Hopf algebra. This will not be a quadratic algebra, because the relations are no longer homogeneous. However, one can introduce a central indeterminate and use it to homogenize the relations, so that a quadratic algebra is obtained (cf. the above example concerning $U(\mathfrak{g})$).

Convention. If M is a graded module over a graded ring, then M^* is defined to be $\bigoplus(M_n^*)$, the sum of the duals of the homogeneous components.

The Koszul complex. Let A be a quadratic algebra, with augmentation ideal $J = \langle A_1 \rangle$, and consider $\mathbb{C} = A/J$ as a left A-module. The algebra $A^!$ may be used to construct a complex of free left A-modules which is a *potential resolution* of the A-module $\mathbb{C}$ (the exactness of the complex is usually difficult to determine). When A is the symmetric algebra on V, this is the Koszul complex. Consider $K(A) = A \otimes (A^!)^*$ as a free left A-module, graded by $K_n(A) = A \otimes (A_n^!)^*$, and as a right $A^!$-module. There is a natural A-module map $K_0(A) \cong A \to \mathbb{C}$, the augmentation. There is a distinguished element of the algebra $A \otimes A^!$, namely $e = \sum x_i \otimes \xi_i$ where x_i is a basis for V, and ξ_i the dual basis for V^*; thus e is the identity element of $V \otimes V^* \cong \mathrm{End}_{\mathbb{C}} V$. The importance of e lies in the fact that, as an element of $A \otimes A^!$, $e^2 = 0$, and hence right multiplication by e defines a differential on $K(A)$. Write d for the differential, and call $(K(A), d)$ the *Koszul complex* of A.

Define the *Hilbert series* of A to be $H(A, t) = \sum(\dim A_n)t^n$.

THEOREM. [Lo] *If A is a quadratic algebra, the following conditions are equivalent:*

(a) *the Koszul complex $(K(A), d)$ is exact;*
(b) *the Yoneda algebra, $\mathrm{Ext}_A^*(\mathbb{C}, \mathbb{C})$ is generated by $\mathrm{Ext}^1(\mathbb{C}, \mathbb{C})$*
(c) *as graded algebras, $\mathrm{Ext}_A^*(\mathbb{C}, \mathbb{C}) \cong A^!$;*
(d) *there is a graded resolution $\ldots \to P_1 \to P_0 \to \mathbb{C} \to 0$ where P_i is a projective module generated by elements of degree i, and the maps are of degree zero.*

If A satisfies the conditions of the theorem, call A a *Koszul algebra*. If A is a Koszul algebra then $H(A, t).H(A^!, -t) = 1$. It would be interesting to know which quadratic algebras described in this article are Koszul algebras.

EXAMPLE: A polynomial ring is a Koszul algebra, as is an exterior algebra.
In fact these are Koszul duals in the sense that they are the Ext-algebras
for each other.

§6. Regular algebras and other quantum spaces

The philosophy of quantum groups involves not just deformations of
groups, but also deformations of affine space, and thus gives rise to non-
commutative algebras which are deformations of polynomial rings. In 1987
Artin and Schelter [AS] defined a class of graded algebras which have many
of the good properties of polynomial rings. It turns out that a number of
the algebras arising in the context of quantum groups fall within the com-
pass of their definition. However, there are also algebras covered by this
definition which are quite different from those that arise within the context
of quantum groups—these other algebras are related to elliptic curves in a
beautiful way.

DEFINITION [AS]: Let $A = A_0 \oplus A_1 \oplus \cdots$ be a connected (i.e., $A_0 = K$)
graded K-algebra, generated by A_1. Suppose that $\dim_K A_1 < \infty$. We say
that A is a d-dimensional *regular graded algebra* if

   (i) $\operatorname{gl dim} A = d < \infty$;
  (ii) there exists $\rho \in \mathbb{R}$ such that $\dim A_n \leq n^\rho$ for all n (that is
         $GK \dim(A) < \infty$);
 (iii) $\operatorname{Ext}^i_A(K, A) = \begin{cases} K & i = d \\ 0 & i \neq d. \end{cases}$

These conditions put strong restrictions on A. For example, if A is commutative, and regular (in this sense!), then A must be a polynomial ring. If
$d = 1$, the only such A is the polynomial ring $K[X]$. If $d = 2$, then A is of
the form $K[X, Y]$ with a single quadratic relation, which (after a suitable
change of variables) is either $XY - YX = Y^2$, or $XY = \lambda YX$ for some
$0 \neq \lambda \in K$. In particular, the quantum plane gives a regular algebra. If
$d = 3$ then things begin to get interesting. There are 13 classes of regular
algebras (for details see [AS]), however two such classes are of particular
interest: these are the "type A" algebras in their terminology. We shall
talk about one one such class.

Fix $(a, b, c) \in \mathbb{P}^2$, and let $A = \mathbb{C}[X, Y, Z]$ with defining relations

$$aX^2 + bYZ + cZY = 0$$
$$aY^2 + bZX + cXZ = 0$$
$$aZ^2 + bXY + cYX = 0.$$

This algebra is very closely related to the subvariety of $\mathbb{P}^2$, E say, defined by the equation $(a^3 + b^3 + c^3)XYZ - abc(X^3 + Y^3 + Z^3) = 0$. Usually E is an elliptic curve; notice that if $(a, b, c) = (0, 1, -1)$, then $E = \mathbb{P}^2$, and A is the polynomial ring. The structure of A is studied in detail in the papers [ATV1] and [ATV2] and we recall some of their results.

Suppose that (a, b, c) is such that E is an elliptic curve. Then A is regular, is a Koszul algebra, and a noetherian domain with the same Hilbert series as the polynomial ring in three variables. The algebra A also determines an automorphism of E. If this automorphism is of finite order, then A is a finite module over its center. If the automorphism is of infinite order, then the center of A is generated by a single homogeneous element of degree 3. Furthermore in this case A has a unique finite dimensional simple module, the trivial one.

Now suppose that $d = 4$. Not all the regular algebras are known for $d = 4$, however there is one class that has been studied to some extent [SS]. This is a family of algebras defined by E. K. Sklyanin [S1], [S2]. Let $(\alpha, \beta, \gamma) \in \mathbb{C}^3$ lie on the surface $\alpha + \beta + \gamma + \alpha\beta\gamma = 0$. Let $A = \mathbb{C}[a, x, y, z]$ with defining relations

$$ax - xa = \alpha(yz + zy) \qquad\qquad xy - yx = az + za$$
$$ay - ya = \beta(xz + zx) \qquad\qquad yz - zy = ax + xa$$
$$az - za = \gamma(xy + yx) \qquad\qquad zx - xz = ay + ya.$$

If $\{\alpha, \beta, \gamma\} \cap \{0, +1, -1\} = \emptyset$, then A is a regular algebra of dimension 4, is a Koszul algebra, and has the same Hilbert series as the polynomial ring $S(\mathbb{C}^4)$. Moreover, it follows from [ATV2] that A is a domain, and is noetherian. For certain values of $\{\alpha, \beta, \gamma\}$, $U_q(sl(2))$ is a quotient of A. If $(\alpha, \beta, \gamma) = (0, 0, 0)$, then $U(sl(2))$ is a quotient of A.

The "type A" 3-dimensional regular algebras above, and these 4-dimensional Sklyanin algebras are special cases of a general construction due to Odesskii and Feigin [OF]. It is quite certain that these will also turn out to be very interesting examples for ring theorists to consider. As yet very little is known about the module/representation theory of these algebras. For example, the finite dimensional simple modules are not yet known. All of these algebras are closely related to elliptic curves—in fact the algebras in [OF] are defined in terms of theta functions. It is already clear that the properties of these algebras are closely related to the geometry of the curve. This geometric connection continues one of the main themes in the study of enveloping algebras, and provides further evidence that many noncommutative algebras can be best studied by using methods from algebraic geometry.

We have deliberately not given a precise definition of a quantum group. Although some authors call any non-commutative, non-cocommutative Hopf algebra a quantum group, this seems rather too general. For example, it includes not only $\mathcal{O}_q(G)$ but also $U_q(\mathfrak{g})$ which is more properly "the enveloping algebra of a quantum Lie algebra". Another possibility is Manin's approach, where one takes just those algebras of the form $\underline{e}(A)$ where A is a quadratic algebra. However, the following examples are not of that form and should by all rights also be included in any reasonable definition. One could insist that one has a deformation of a genuine Hopf algebra $\mathcal{O}(G)$, but this would exclude some of Manin's examples. It seems quite likely that there still remains a large number of "quantum groups" to be discovered.

Dipper and Donkin's quantum groups. Consider the algebra $D = \mathbb{C}[a, b, c, d]$ defined by relations

$$
\begin{array}{lll}
ab = ba & ac = qca & cb = qbc \\
bd = qdb & dc = cd & ad - da = (1 - q)bc
\end{array}
$$

Then D is a bialgebra, with the "same" coproduct and co-unit as $\mathcal{O}(M_2(\mathbb{C}))$, namely $\Delta(X_{ij}) = \sum_k X_{ik} \otimes X_{kj}$, and $\varepsilon(X_{ij}) = \delta_{ij}$. The center of D is $\mathbb{C}$ (remember q is not a unit). The element $ad - bc$ is a normal element in D, i.e., $(ad - bc)D = D(ad - bc)$. In particular, notice that there is no quantum $SL(2)$ associated to D, because factoring out $\langle ad - bc - 1 \rangle$ gives the coordinate ring of a torus. However, there is a quantum $GL(2)$, namely $D[(ad - bc)^{-1}]$ which has a Hopf algebra structure. The algebra D cannot be obtained from the R-matrix construction of §4. Nevertheless it seems reasonable to consider this algebra as the coordinate ring of (some) quantum 2×2 matrices.

Dipper and Donkin [DD] use D to define a quantum $GL(n)$ which is different from the $\mathcal{O}_q(GL(n))$ defined earlier. Consider $\mathbb{C}[Y_{ij} \mid 1 \le i, j \le n]$ with defining relations such that the map

$$
\begin{bmatrix} a & d \\ c & d \end{bmatrix} \longrightarrow \begin{bmatrix} Y_{ik} & Y_{im} \\ Y_{jk} & Y_{jm} \end{bmatrix}
$$

is an isomorphism from D to $\mathbb{C}[Y_{ik}, Y_{im}, Y_{jk}, Y_{jm}]$ whenever $i < j$ and $k < m$. Now $\mathbb{C}[Y_{ij}]$ is made into a bialgebra with coproduct $\Delta(Y_{ij}) = \sum_{k=1}^{n} Y_{ik} \otimes Y_{kj}$ and co-unit $\varepsilon(Y_{ij}) = \delta_{ij}$. There is a homogeneous normalising element of degree n, which is group-like; this element is the quantum determinant, and can be inverted to obtain a new quantum $GL(n)$. The basic ring theoretic properties of this algebra are the same as those of $\mathcal{O}_q(GL(n))$.

More recently Takeuchi [T3] has constructed a two parameter family of deformations of $\mathcal{O}(GL(n))$ which includes both the Dipper-Donkin examples, and the "official" $\mathcal{O}_q(GL(n))$ described earlier.

EXAMPLE: ([JGW]). Another quantum group "modelled" on $U(sl(2))$ is the algebra $\mathbb{C}[X, Y, K, K^{-1}, \xi, \xi^{-1}]$ with relations

$$KXK^{-1} = qX \qquad KYK^{-1} = q^{-1}Y \qquad \xi X \xi^{-1} = -q^{-1}X \qquad \xi Y \xi^{-1} = -qY$$

$$K\xi = \xi K \qquad XY - YX = \frac{K\xi - K^{-1}\xi^{-1}}{q - q^{-1}} \qquad X^2 = Y^2 = 0$$

The Hopf structure is given by

$$\begin{aligned} \Delta(K) &= K \otimes K & s(K) &= K^{-1} & \varepsilon(K) &= 1 \\ \Delta(\xi) &= \xi \otimes \xi & s(\xi) &= \xi^{-1} & \varepsilon(\xi) &= 1 \\ \Delta(X) &= X \otimes K + \xi^{-1} \otimes X & s(X) &= -\xi X K^{-1} & \varepsilon(X) &= 0 \\ \Delta(Y) &= Y \otimes \xi + K^{-1} \otimes Y & s(Y) &= -KY\xi^{-1} & \varepsilon(Y) &= 0 \end{aligned}$$

This algebra is of GK-dimension 2, and is a finite module over its centre $\mathbb{C}[XY + YX, (K\xi)^2, (K\xi)^{-2}]$. The subalgebra generated by X and Y is the enveloping algebra of a Lie superalgebra (with X and Y odd), and the algebra itself is a quotient of a smash product of this subalgebra with $\mathbb{C}[K, K^{-1}, \xi, \xi^{-1}]$. The basic properties of this algebra, and its representations have been worked out in [JGW].

§7. Woronowicz's point of view

At about the same time as $U_q(\mathfrak{g})$ was invented by Drinfeld and Jimbo, Woronowicz invented some C^*-algebras which turned out to be closely related. We will mainly discuss the simplest case, corresponding to $sl(2)$.

First recall that if G is a compact Lie group, then the representation theory of G can be studied through a Hopf algebra structure on the C^*-algebra, $C(G)$, of continuous $\mathbb{C}$-valued functions on G. Let $G = SU(2)$ be the compact, real, Lie group of unitary 2×2 matrices with determinant 1, $\begin{bmatrix} a & -\bar{b} \\ b & \bar{a} \end{bmatrix}$ where $a, b \in \mathbb{C}$ and $a\bar{a} + b\bar{b} = 1$. The condition that $a\bar{a} + b\bar{b} = 1$ is equivalent to the requirement that

$$\begin{bmatrix} a & -\bar{b} \\ b & \bar{a} \end{bmatrix} \begin{bmatrix} \bar{a} & \bar{b} \\ -b & a \end{bmatrix} = \begin{bmatrix} 1 & 0 \\ 0 & 1 \end{bmatrix}$$

Thus $C(G)$ is the C^*-algebra generated (as a C^*-algebra) by functions α, γ where α gives the 11-entry and γ gives the 21-entry. We can make $C(G)$ a Hopf C^*-algebra (Warning: the definition of a Hopf C^*-algebra is not exactly the same as that of a Hopf-algebra since the ordinary tensor product of a C^*-algebra is not a C^*-algebra; it must be completed). The comultiplication is given by

$$\Delta(\alpha) = \alpha \otimes \alpha - \gamma^* \otimes \gamma \qquad \Delta(\gamma) = \gamma \otimes \alpha + \alpha^* \otimes \gamma.$$

This is the restriction of the coproduct for $GL(2)$, since $-\gamma^*$ gives the 12-entry and α^* gives the 22-entry. Similarly, $s(\alpha) = \alpha^*$, $s(\gamma) = -\gamma$, $\varepsilon(\alpha) = 1$ and $\varepsilon(\gamma) = 0$.

For $\nu \in [-1, 1]$, Woronowicz defines a (non-commutative) C^*-algebra $C(S_\nu U(2))$, generated (as a C^*-algebra) by elements α and γ subject to relations

$$\alpha^* \alpha + \gamma^* \gamma = 1 \qquad\qquad \alpha\alpha^* + \nu^2 \gamma = 1$$
$$\gamma\gamma^* = \gamma^* \gamma \qquad\qquad\qquad \alpha\alpha^* = \alpha^* \alpha$$
$$\alpha\gamma = \nu\gamma\alpha \qquad\qquad\qquad \alpha\gamma^* = \nu\gamma^*\alpha.$$

These can be expressed succinctly by saying that

$$\begin{bmatrix} \alpha & -\nu\gamma^* \\ \gamma & \alpha^* \end{bmatrix} \begin{bmatrix} \alpha^* & \gamma \\ -\nu\gamma & \alpha \end{bmatrix} = \begin{bmatrix} 1 & 0 \\ 0 & 1 \end{bmatrix}.$$

The Hopf algebra structure is given by maps Δ, ε and s, as follows:

$$\Delta(\alpha) = \alpha \otimes \alpha - \nu\gamma^* \otimes \gamma \qquad \Delta(\gamma) = \gamma \otimes \alpha + \alpha^* \otimes \gamma$$
$$s(\alpha) = \alpha^* \quad s(\alpha^*) = \alpha \qquad s(\gamma) = \nu\gamma \qquad\qquad s(\gamma^*) = -\nu^{-1}\gamma^*$$
$$\varepsilon(\alpha) = \varepsilon(\alpha^*) = 1 \qquad\qquad \varepsilon(\gamma) = \varepsilon(\gamma^*) = 0.$$

If $\nu = 1$, then one obtains $C(SU(2))$, so $C(S_\nu U(2))$ is the C^*-algebra of deformation of $SU(2)$.

Set $q^2 = \nu$. There is a $*$-algebra homomorphism $\varphi : \mathbb{C}[\alpha, \alpha^*, \gamma, \gamma^*] \to \mathcal{O}_q(SL(2))$ given by $\varphi(\alpha) = a$, $\varphi(\alpha^*) = d$, $\varphi(\gamma) = q^{-1}c$, and $\varphi(\gamma^*) = -q^{-1}b$. This is also a homomorphism of Hopf algebras. Thus $\mathbb{C}[\alpha, \alpha^*, \gamma, \gamma^*] \cong \mathcal{O}_q(SL(2))$ as $*$-Hopf algebras, so $C(S_\nu U(2))$ is a C^*-completion of $\mathcal{O}_q(SL(2))$.

Woronowicz obtains an analogue of the Lie algebra of a Lie group by considering functions on A coming from the coordinate functions on the finite dimensional representations of A. He thus obtains the appropriate analogue of (left) invariant differential operators on $S_\nu U(2)$. These generate an algebra $W_\nu = \mathbb{C}[\nabla_0, \nabla_1, \nabla_2]$ [W, Table 7, p. 150] which in retrospect is isomorphic to a subalgebra of $U_q(sl(2))$. It is defined by the relations

$$\nu \nabla_2 \nabla_0 - \nu^{-1} \nabla_0 \nabla_2 = \nabla_1$$
$$\nu^2 \nabla_1 \nabla_0 - \nu^{-2} \nabla_0 \nabla_1 = (1 + \nu^2) \nabla_0$$
$$\nu^2 \nabla_2 \nabla_1 - \nu^{-2} \nabla_1 \nabla_2 = (1 + \nu^2) \nabla_2.$$

If $\nu = q^2$, there is an injective algebra homomorphism $W_\nu \to U_q(sl(2))$ defined by $\nabla_0 \mapsto -qFK$, $\nabla_1 \mapsto qEK$, $\nabla_2 \mapsto (K^4 - 1)/(q^{-4} - 1)$. Identify W_ν with its image in $U_q(sl(2))$. Woronowicz proves that W_ν has exactly one simple module in each finite dimension, except that there is a 1-parameter family of 1-dimensional modules [W, Theorem 5.4]. If we localise a little the situation is even nicer: $W_\nu[K^{-4}]$ has exactly 1 simple module in each finite dimension, and every finite dimensional $W_\nu[K^{-4}]$-module is semisimple.

Consider the inclusions

$$U_q(sl(2)) = \mathbb{C}[E, F, K^{\pm 1}]$$
$$\supset \mathbb{C}[EK, FK, K^{\pm 2}]$$
$$\supset \mathbb{C}[EK, FK, K^{\pm 4}]$$
$$= W_\nu[K^{-4}].$$

The first two are Hopf algebras, but the last one is not. If $n > 0$, then $U_q(sl(2))$ has 4 distinct n-dimensional simple modules, $W_q(sl(2))$ has 2 distinct n-dimensional simple modules. In terms of finite dimensional simple modules $\mathbb{C}[EK, FK, K^{\pm 4}]$ is the most like $U(sl(2))$, having one in each dimension, and no non-split extensions.

Recall the action of $U_q(sl(2))$ on $\mathcal{O}_q(\mathbb{C}^2)$. The restriction of this to W_ν has a nice interpretation in terms of skew derivations (see [MS] for details). The action of W_ν is such that K^4 acts as the automorphism $X \mapsto q^4 X, Y \mapsto q^{-4} Y$ while both EK and FK act as the K^2-derivations

$$EK : X \mapsto 0, Y \mapsto X \qquad \text{and} \qquad FK : X \mapsto Y, Y \mapsto 0.$$

Under this action, the nth homogeneous component of $\mathcal{O}_q(\mathbb{C}^2)$ becomes the unique $(n + 1)$-dimensional simple W_ν-module.

Woronowicz [W2] also defines analogues of the groups $SU(n)$ for all n. Consider a C^*-algebra with a dense subalgebra $\mathfrak{G} = \mathbb{C}[u_{ij} | 1 \leq i, j \leq n]$ with relations amongst the u_{ij} such that

(a) there is a homomorphism of algebras $\Delta : \mathfrak{G} \to \mathfrak{G} \otimes \mathfrak{G}$ such that
$\Delta(u_{ij}) = \sum_{k=1}^{n} u_{ik} \otimes u_{kj}$

(b) there is an anti-automorphism $s : \mathfrak{G} \to \mathfrak{G}$ such that $s(s(a^*)) = a$ for all $a \in \mathfrak{G}$;

(c) there is an algebra homomorphism $\varepsilon : \mathfrak{G} \to \mathbb{C}$ such that $\varepsilon(u_{ij}) = \delta_{ij}$.

(d) In $M_n(\mathfrak{G})$ the matrix $s(u_{ij})$ is the inverse of (u_{ij}).

§8. q a root of unity

The preceeding sections have ignored what happens when q is a root of unity. In part this is because matters change so dramatically in that case, and finite dimensional $U_q(\mathfrak{g})$-modules are no longer semisimple. For example, $U_q(sl(2))$ is a finite module over its center. It is quite easy to see that the same is true of $\mathcal{O}_q(\mathbb{C}^n)$, and Parshall and Wang [PW] show that this is also the case for $\mathcal{O}_q(M_n(\mathbb{C}))$. Is the same true of $U_q(\mathfrak{g})$, and the algebras arising from quantum $SP(2n)$ and $SO(n)$?

It would be interesting to describe all the finite dimensional simple modules over $U_q(sl(2))$ for q a root of unity, and compare this with the classification of finite dimensional simple modules over $U(sl(2))$ in the finite characteristic case [RS]. One could also try to do this for a general $U_q(\mathfrak{g})$.

Lusztig suggests [L2] that it is not so much the representation theory of $U_q(\mathfrak{g})$ as that of a certain subalgebra (analogous to the Kostant $\mathbb{Z}$-form) which is of interest. In [L3] Lusztig shows that there is a finite dimensional quotient of this subalgebra which is a Hopf algebra. He conjectures that there is a close relationship between the characters of the simple modules over this finite dimensional algebra when q is a pth root of 1, and the characters of the finite dimensional simple modules over the corresponding algebraic group defined over $\overline{\mathbb{F}}_p$.

The case of $U_q(sl(2))$ with $q^r = 1$ is discussed in [RT]. In this case $U_q(sl(2))/\langle E^r, F^r \rangle$ is a finite dimensional Hopf algebra. It is not semisimple, it has $r - 1$ simple modules, and these have dimensions $1, 2, \dots, r - 1$. Given a link ($\equiv$ a closed braid) one defines a certain module over this finite dimensional algebra. The crossings of the strands in the link determine an endomorphism of this module (built up in part from the R-matrix for $sl(2)$). This endomorphism turns out to be an invariant of the link. It would probably be interesting to study these algebras in the spirit of Auslander-Reiten. For example, what are their indecomposable representations?

§9. What next for ring theorists?

This section is (of necessity) somewhat speculative.

New examples of non-commutative rings which are "natural", and have a rich structure, appear only intermittently. For the past 15 years enveloping algebras of Lie algebras, and rings of differential operators have been the newer examples to influence the development of non-commutative ring theory. I believe that quantum groups, and quadratic algebras are the next class of examples which will influence its direction. To some extent, non-commutative ring theorists have been studying quantum spaces all along; for example, Drinfeld, Manin and Parshall-Wang define the category of quantum spaces to be the dual category of the category of K-algebras!

So what are some of the questions which one might pursue?

"Clearly", $U_q(\mathfrak{g})$ is just like $U(\mathfrak{g})$. Prove this. There should be a classification of primitive ideals similar to the classical case (e.g., is there a version of Duflo's Theorem?). In fact, as for the finite dimensional representations, one expects the same classification (except for the annoying problem of $4^{\mathrm{rank}\mathfrak{g}}$ 1-dimensional modules). On the one hand it is somewhat unexciting that the same result is obtained (for the finite dimensional representations), but at least satisfying to know that the enveloping algebra techniques are robust enough to cope with a slightly different algebra. Is there some precise relationship (a functor?) between the categories of $U(\mathfrak{g})$ and $U_q(\mathfrak{g})$-modules? Compare this to what happens when q is a root of unity in the paper [Xi].

One would also expect that the behavior of $U_q(\mathfrak{n}^+)$ and $U_q(\mathfrak{b})$ should be analogous to the classical case. For example, are primitive ideals of $U_q(\mathfrak{n}^+)$ maximal, and are the primitive quotients q-analogues of the Weyl algebra, when q is not a root of unity?

Many questions are given at the end of Manin's notes [M2]. Of those questions, perhaps the most interesting involve rings which are the analogues of the coordinate rings of homogeneous spaces. However, note that in terms of the questions ring theorists have tended to ask, $\mathcal{O}_q(\mathbb{C}^n)$ is actually rather easy to understand. The coordinate rings of quantum Euclidean and symplectic space should also be studied. Are they Koszul algebras? Are they noetherian? What are their primitive ideals, and homogeneous prime ideals?

Perhaps some new questions, having more geometric and representation theoretic origin, will be interesting. For example, find the analogue of G/B, and in particular its homogeneous coordinate ring, $S_q(G/B)$ say. Is there a Borel-Weil-Bott theorem? What is the analogue of a line bundle? Is there a bijection between closed points of $\mathrm{Proj}S_q(G/B)$ and elements of the Weyl group W? After the quantum $\mathbb{P}^n$ example, one might expect a quantum G/B with basis of open sets being the W-orbit of the large Bruhat cell Bw_0B/B.

The quadratic algebras of [ATV], Feigin and Odesskii, and Sklyanin all deserve further study. Classify their finite dimensional simple modules, and the primitive ideals; one should try to do this in geometric terms along the lines of the Dixmier correspondence for $U(\mathfrak{g})$ when $\mathfrak{g}$ is solvable.

Do the results and techniques (in [ATV1], [ATV2]) for the low dimensional cases extend to higher dimensional cases? For example, look at the algebras that can be constructed from the data $(X, \sigma, \mathcal{L})$ where X is a smooth projective variety, σ an automorphism, and $\mathcal{L}$ an ample line bundle. For example, when $X = \mathbb{P}^n$, and $\mathcal{L} = \mathcal{O}(1)$ then the algebra $B = \bigoplus_{n=0}^{\infty} H^0(X, \mathcal{L} \otimes \mathcal{L}^{\sigma} \otimes \cdots \otimes \mathcal{L}^{\sigma^n})$ defined in [ATV1] and [OF] is a Koszul algebra, and an iterated Ore extension on $n+1$ generators, and it should be possible to understand its primitive spectrum using inductive arguments. In particular, is there a bijection between its primitive ideals and a set of quasi-affine varieties in $\mathbb{C}^{n+1}$, such that (i) $\mathbb{C}^{n+1}$ is the disjoint union of the quasi-affine varieties, (ii) the GK-dimension of B/J equals the dimension of the corresponding variety, and (iii) there is an inclusion of primitive ideals $I \subset J$ if and only if the reverse inclusion holds for the closures of the corresponding varieties. The same question should be asked for the coordinate rings of the quantum homogeneous spaces e.g., for the "easy" cases of $\mathcal{O}_q(Sp\mathbb{C}^{2n})$ and $\mathcal{O}_q(o\mathbb{C}^n)$.

Many of the algebras discussed in this survey (and enveloping algebras) are deformations of polynomial rings. Let A be a poynomial ring, $P \in \mathrm{Der}\, A \otimes_{\mathbb{C}} \mathrm{Der} A, t \in \mathbb{C}$, and $\mu : A \otimes A \to A$ the multiplication. Write $P(a, b)$ for $P(a \otimes b)$. Define a new multiplication, $*$, on A by

$$a * b = ab + t\mu P(a, b) + \frac{t^2}{2!}\mu P^2(a, b) + \cdots + \frac{t^n}{n!}\mu P^n(a, b) + \cdots .$$

Find conditions on P such that $*$ is associative, and consider the new algebra, A_t say, as a deformation of A, with $A_0 \cong A$. Describe the primitive ideals of A_t in terms of those of A and their behaviour with respect to the Poisson bracket $\{a, b\} = P(a, b) - P(b, a)$. The idea is that it may be possible to understand from a single point of view a large class of algebras that at present all appear to require different analysis. In particular, all the algebras we have discussed are "deformations" of polynomial rings; can this be formalised in such a way that one can then transfer the study of the non-commutative algebra to the study of the commutative algebra with some extra structure? For example, it is quite easy to see that $U(\mathfrak{g})$ is a deformation of a polynomial ring in the sense of Gerstenhaber [G]; in fact it is so easy because there are uncountably many ways of doing it. There should be one "right way", which will be particularly illuminating. For example, the symmetrisation map is sometimes the "right" deformation. In [GS], $\mathcal{O}_q\big(GL(n)\big)$ is described as a deformation of $\mathcal{O}\big(GL(n)\big)$. It is shown there that $\mathcal{O}_q\big(M_n(\mathbb{C})\big)$ can be described as $\mathcal{O}\big(M_n(\mathbb{C})\big)$ with a new multiplication $*$ in such a way, that the quantum determinant equals the classical determinant.

Understand the basic properties of the rings $\mathcal{O}_q\big(SP(2n)\big)$ and $\mathcal{O}_q\big(SO(n)\big)$.

Find conditions on the matrix R such that $\mathcal{O}_R(\mathbb{C}^n)$ (resp. $\mathcal{O}_R\big(M_n(\mathbb{C})\big)$) have the same growth as a polynomial ring in n (resp. n^2) indeterminates.

Construct some algebras $\underline{e}(A)$ using Manin's machine for various "nice" quadratic algebras A. For example, Holland has asked if Manin's machine yields $\mathcal{O}_q\big(SP(2n)\big)$ and $\mathcal{O}_q\big(SO(n)\big)$ when $\mathcal{O}_q(Sp\mathbb{C}^{2n})$ and $\mathcal{O}_q(o\mathbb{C}^n)$ are inserted. Another interesting case is to take the commutative ring $A = \mathbb{C}[X, Y, Z]/\langle X^2 + Y^2 + Z^2 \rangle$, construct $\underline{e}(A)$ and study its finite dimensional simple modules and comodules. This may be similar to $\underline{e}\big(\mathcal{O}(\mathbb{C}^2)\big)$ since $\mathrm{Proj}A \cong \mathrm{Proj}\mathcal{O}(\mathbb{C}^2) \cong \mathbb{P}^1$.

Given a Hopf algebra, H say, constructed by Manin's machine, is there a "nice" Hopf algebra H' such that the category of finite dimensional comodules over H is equivalent to the dual of the category of finite dimensional simple H'-modules. By "nice" I mean (at least) that if H is of finite GK-dimension, then H' has the same GK-dimension (so I do not take the full cofinite dual). In [FRT] there is a general procedure which for a Hopf algebra obtained from $\mathcal{O}_R\big(M_n(\mathbb{C})\big)$ (e.g., an $\mathcal{O}_q(G)$), constructs a Hopf dual which gives in particular the $U_q(\mathfrak{g})$. The dual algebras obtained from other R than those given in §4 may also give interesting examples. In addition, is there a nice dual of Dipper and Donkin's quantum $GL(2)$? The idea is that if the $U_q(\mathfrak{g})$ are so like $U(\mathfrak{g})$, then perhaps one should look for examples which are more novel. In addition, it is easier to study modules than comodules.

It is commonly believed that we do not yet know the "right" class of non-commutative rings which will have a theory "like" that of commutative rings. The new examples (such as quantum groups) are important in guiding us towards the right class.

REFERENCES

[An] H. H. Andersen, *The Linkage Principle and the sum formula for quantum groups.* Preprint, 1989.

[AnP] H. H. Anderson and P. Polo, *Representations of quantum algebras.* Preprint, 1990.

[AS] M. Artin and W. Schelter, *Graded algebras of global dimension 3,*, Adv. Math. **66** (1987), 171–216.

[ATV1] M. Artin, J. Tate and M. van den Bergh, *Some algebras associated to automorphisms of curves.* Preprint, 1989.

[ATV2] ⸻, *Modules over regular algebras of dimension 3.* Preprint, 1989.

[Av] M. Artin and M. van den Bergh, *Twisted Homogeneous Coordinate Rings.* Preprint, 1990.

[BF] J. Backelin and R. Froberg, *Koszul algebras, Veronese subrings and rings with linear resolution,* Revue Roumaine de Math. Pures et Appl. **30** (1985), 85–97.

[B] R. J. Baxter, "Exactly solved models in statistical mechanics," Academic Press, New York, 1982.

[BG] A. Beilinson and V. Ginsburg, *Mixed categories, Ext-duality and Representations (results and conjectures).* Preprint.

[Be] A. A. Belavin, *Discrete Groups and the Integrability of Quantum Systems,* Func. Anal. and its Appl. **14** (1980), 260–267.

[BD1] A. A. Belavin and V. G. Drinfeld, *On the solutions of the classical Yang- Baxter equations,* Func. Anal. and Appl. **16** (1982), 159–180.

[BD2] ⸻, *Triangle Equations and simple Lie algebras,* Sov. Sci. Rev. **C4** (1984).

[Berg] G. Bergman, *Diamond Lemma for Ring Theory,* Adv. Math. **29** (1978), 178–218.

[Br] K. Bragiel, *Twisted SU(3) group.* to appear.

[C] I. V. Cherednik, *Some finite dimensional representations of generalized Sklyanin algebras,* Func. Anal. and Appl **19** (1985), 77–79..

[DD] R. Dipper and S. Donkin, *Quantum GL_n.* Preprint.

[DR] P. Doubilet and G. C. Rota, *Skew-symmetric Invariant Theory,* Adv. Math. **21** (1976), 196–203.

[D1] V. G. Drinfeld, *Hamiltonian structures on Lie groups, Lie bialgebras, and the geometric meaning of the Yang-Baxter equations,* Sov. Math. Dokl. **32** (1985), 254–258.

[D2] ⸻, *Hopf algebras and the quantum Yang-Baxter equation,* Sov. Math. Dokl. **32** (1985), 254–258.

[D3] ⸻, *Degenerate affine Hecke algebras and Yangians,* Func. Anal. and Appl. **20** (1986), 58–60.

[D4] ⸻, *Quantum Groups,* Proc. Int. Cong. Math.; Berkeley, **1** (1986), 798–820.

[D5] ⸻, *A new realisation of Yangians and quantised affine algebras,* Sov. Math. Dokl. **36/296** (1988), 212–216.

[D6] ⸻, *On quadratic commutation relations in the quasi-classic limit,* Mat. Fizika Funkc. Analiz. Kiev, 25–33. (in Russian).

[DPW] J. Du, B. Parshall and J-P. Wang, *Two parameter quantum linear groups and the Hyperbolic invariance of q-Schur algebra.* Preprint, 1990.

[FRT] L. D. Faddeev, N. Y. Reshetikhin and L. A. Takhtajan, *Quantization of Lie groups and Lie algebras.* Preprint LOMI.

[FT] L. D. Faddeev, and L. A. Takhtajan, *Liouville model on the lattice*, in "Lecture Notes in Physics, No. 246," Springer-Verlag, 1986, pp. 166–178.

[G] M. Gerstenhaber, *On the deformation of rings and algebras*, Ann. Math. **79** (1964), 59–103.

[GS] M. Gerstenhaber and S. D. Schack, *Quantum groups as deformations of Hopf algebras*, Proc. Nat. Acad. Sci. **87** (1990), 478–481.

[H] T. Hayashi, *Q-analogues of Clifford and Weyl algebras. Spinor and oscillator representations of quantum enveloping algebras.* Preprint, Nagoya, 1990.

[J1] M. Jimbo, *A q-difference analogue of $U(\mathfrak{g})$ and the Yang-Baxter equation*, Lett. Mat. Phys. **10** (1985), 63–69.

[J2] ———, *Quantum R matrix for the generalised Toda system*, Commun. Math. Phys. **102** (1986), 537–547.

[J3] ———, *A q-analogue of $U(\mathfrak{gl}(n+1))$, Hecke algebra and the Yang-Baxter equation*, Lett. Math. Phys. **11** (1986), 247–252.

[JGW] Naihuan Jing, Mo-Lin Ge, Yong-Shi Wu, *New quantum group associated with a "non-standard" Braid group representation.* Preprint, 1990.

[Jo1] V. F. R. Jones, *Polynomial invariants of knots via von Neumann algebras*, Bull. Amer. Math. Soc. **12** (1985), 103–111.

[Jo2] ——————, *Hecke algebra representations of braid groups, and link polynomials*, Ann. Math. **126** (1987), 335–388.

[Jo3] ——————, *On knot invariants related to some statistical mechanical models.*

[KR1] A. A. Kirillov and N. Yu. Reshetikhin, *The Yangians, Bethe Ansatz and combinatorics*, Lett. Math. Phys. **12** (1986), 199–208.

[KR2] ——————————, *Representations of the algebra $U_q(sl(2))$, q-orthogonal polynomials and invariants of links.* LOMI preprint (1988).

[Ko] T. Kohno, *Monodromy representations of Braid groups and Yang-Baxter equations*, Ann. Inst. Fourier **37** (1987), 139–160.

[KS] Y. Kosmann-Schwarzbach, *Poisson-Drinfeld groups;* Proc. Oberwolfach Conf. on non-linear evolution equations ; M. Albowitz, B. Fuchsteiner, M. Kruskal ed. World Scientific Publ. (1986).

[KR] P. P. Kulish and N. Reshetikhin, *Quantum linear problem for the Sine-Gordon equation and higher representations*, J. Sov. Math. **23** (1983), 2435–2441.

[KRS] P. P. Kulish, N. Reshetikhin, and E. K. Sklyanin, *Yang-Baxter equation and representation theory*, Lett. Math. Phys. **5** (1981), 393–403..

[KS] P. P. Kulish and E. K. Sklyanin, *Solutions of the Yang-Baxter equation*, J. Sov. Math. **19** (1982), 1596–1620.

[Lo] C. Lofwall, *On the subalgebra generated by the 1-dimensional elements in the Yoneda Ext-algebra*, Springer LNM **1183** (1988), 291–338.

[L1] G. Lusztig, *Quantum deformations of certain simple modules over enveloping algebras*, Adv. Math. **70** (1988), 237–249.

[L2] ———, *Modular representations and quantum groups*, Contemp. Math. **82** (1989), 59–78.

[L3] ———, *Finite dimensional Hopf algebras arising from quantum groups.* Preprint.

[L4] ———, *Quantum groups at roots of 1.* Preprint.

[L5] ———, *Canonical bases arising from quantized enveloping algebras.* Preprint, 1990.

[Maj1] S. Majid, *Quasitriangular Hopf Algebras and Yang-Baxter equations*, Inter. J. Mod. Phys. **5** (1990), 1–91.

[Maj2] S. Majid and Ya. S. Soibelman, *Rank of quantized enveloping algebras and modular functions.* Preprint, 1990.

[M1] Yu. I. Manin, *Some remarks on Koszul algebras and Quantum groups*, Ann. Inst. Fourier **37** (1987), 191–205.

[M2] —————, "Quantum groups and Non-commutative geometry," Les Publ. du Centre de Récherches Math., Universite de Montreal, 1988.

[MM] T. Masuda, K. Mimachi, Y. Nakagami, M. Noumi, M. Ueno, *Representations of Quantum groups and a q-analogue of orthogonal polynomials*, C. R. Acad. Sci. Paris **307** (1988), 559–564.

[MP] J. C. McConnell and J. J. Pettit, *Crossed products and multiplicative analogues of Weyl algebras*, J. Lond. Math. Soc. **38** (1988), 47–55.

[MS] S. Montgomery and S. P. Smith, *Skew derivations and $U_q(sl(2))$*, Israel J. Math.. to appear.

[NYM] M. Noumi, H. Yamada and K. Mimachi, *Finite dimensional representations of the quantum group $GL_q(n+1, \mathbb{C})$ and the zonal spherical functions on $U_q(n)$ \ $U_q(n+1)$*. Preprint, 1989.

[OF1] A. V. Odesskii and B. L. Feigin, *Sklyanin algebras associated with an elliptic curve*. Preprint (1988).

[OF2] —————, *Elliptic Sklyanin Algebras*, Func. Anal. Appl. **23** (1989), 45–54.

[O] G. I. Olshanskii, *Yangians and universal enveloping algebras*, LOMI 164 (1987), 142–150.

[PW1] B. Parshall and J-P. Wang, *Quantum Linear Groups I*. Preprint. University of Virginia (1989).

[PW2] —————, *Quantum Linear Groups II*. Preprint.University of Virginia (1989).

[Po] P. Podlcs,, *Quantum Spheres*, Lett. Math. Phys. **14** (1987), 193–202.

[Pr] S. B. Priddy, *Koszul resolutions*, Trans. Amer. Math. Soc. **152** (1970), 39–60.

[Re1] N. Reshetikhin, Theoret. Math. Phys. **63** (1985).

[Re2] —————, *Quantized universal enveloping algebras, the Yang-Baxter equation and invariants of links I.*, LOMI (1988,), E-4-87. Preprint. *II.*, LOMI (1988), E-17-87. Preprint.

[RT] N. Reshetikhin and V. G. Turaev, *Invariants of 3-manifolds via link polynomials and quantum groups*, MSRI (1989). Preprint.

[Ri] C. M. Ringel, *Hall Algebras and Quantum Groups*. Preprint. Bielefeld (1989).

[R1] M. Rosso, *Comparaison des groupes $SU(2)$ quantiques de Drinfeld et de Woronowicz*, C. R. Acad. Sci. Paris **304** (1987), 323–326.

[R2] —————, *Representations irreducibles de dimension finie du q-analogue de l' algebre enveloppante d'une algebre de Lie semisimple*, C. R. Acad. Sci. Paris **305** (1987), 587–590.

[R3] —————, *Finite dimensional representations of the quantum analog of the enveloping algebra of a complex semisimple Lie algebra*, Comm. Math. Phys. **117** (1988), 581–593.

[R4] —————, *Groupes quantiques et modeles à vertex de V. Jones en theorie des noeuds*, C. R. Acad.Sci. Paris **307** (1988), 207–210.

[R5] —————, *An analogue of the PBW theorem and the universal R-matrix for $U_h(sl(n+1))$*. Preprint. Palaiseau (1989).

[R6] —————, *Analogues de la forme de Killing et du théorème d'Harish-Chandra pour les groupes quantiques*. Preprint. Palaiseau (1989).

[RS] A. N. Rudakov and I. R. Shafarevich, *Irreducible representations of a simple three dimensional Lie algebra over a field of finite characteristic*, Math. Notes **2** (1968), 760–767.

[S1] E. K. Sklyanin, *Some algebraic structures connected with the Yang-Baxter equation*, Func. Anal. and Appl. **16** (1982), 263–270.

[S2] —————, *Some algebraic structures connected with the Yang-Baxter equation. Representations of quantum algebras*, Func. Anal. and Appl. **17** (1983), 273–284.

[S3] —————, *An algebra generated by quadratic relations*, Usp. Mat. Nauk. **40** (1985), 214.

[SS] S. P. Smith and J. T. Stafford, *Regularity of the 4-dimensional Sklyanin algebra.* in preparation.

[So] Ya. S. Soibelman, *Irreducible representations of the functional algebra of quantised SU(n) and the Schubert cells*, Func. Anal. and Appl. (to appear).

[TT] E. Taft and J. Towber, *Quantum deformation of flag schemes and Grassman schemes I, a q-deformation of the shape algebra for GL(n)*. Preprint (1989).

[T1] M. Takeuchi, *Quantum Orthogonal and Symplectic Groups and their embedding into Quantum GL(n)*, Proc. Japan Acad. **65** (1989), 55–58.

[T2] —————, *Matrix Bialgebras and Quantum Groups*. Preprint. Tsukuba (1989).

[T3] —————, *A two parameter quanitization of GL(n)*; Summary.

[Tak1] L. A. Takhtajan, *Quantum Groups and Integrable Models*. Preprint. Steklov Inst. (1988).

[Tak2] —————, *Noncommutative homology of quantum tori*, Func. Anal. Appl. **23** (1989), 147–149.

[Tan1] T. Tanisaki, *Harish-Chandra isomorphisms for Quantum algebras*. Preprint, Osaka Univ. (1989).

[Tan2] —————, *Finite dimensional representations of quantum groups*. Preprint, Osaka Univ. (1989).

[Tu] V. G. Turaev, *The Yang-Baxter equation and invariants of links*, Invent. Math. **92** (1988), 527–553.

[VS] L. L. Vaksman and Ya. S. Soibelman, *The algebra of functions on quantised SU(2)*, Func. Anal. and Appl. **22** (1988), 170–181.

[V] J-L. Verdier, *Groupes Quantiques*, Seminaire Bourbaki (1986–87,). No. 685.

[We] H. Wenzl, *Braid group representations and the quantum Yang-Baxter equation.* Preprint.

[W1] S. L. Woronowicz, *Twisted SU(2)-group. An example of a non-commutative differential calculus*, Publ. R.I.M.S., Kyoto Univ. **23** (1987), 117–181.

[W2] —————, *Compact Matrix pseudogroups*, Comm. Math. Phys. **111** (1987), 613–615.

[W3] —————, *Tannaka-Krein duality for compact Matrix pseudogroups*, Invent. Math. **93** (1988), 35–76.

[X] Xi, Nanhua, *Representations of Finite Dimensional Hopf algebras arising from Quantum Groups*. Preprint (1989).

[Y1] H. Yamane, *A Poincare-Birkhoff-Witt Theorem for the quantum group of type A_n*, Proc. Japan Acad. **64** (1988), 385–386.

[Y2] —————, *A PBW Theorem for quantized universal enveloping algebras of type A_N*, Publ. RIMS Kyoto **25** (1989), 503–520.

Department of Mathematics, University of Washington, Seattle, WA 98195